LAND RENE'

Reworking the Coun

Peter Hetherington

BRISTOL
UNIVERSITY
PRESS

First published in Great Britain in 2021 by

Bristol University Press
University of Bristol
1-9 Old Park Hill
Bristol
BS2 8BB
UK
t: +44 (0)117 954 5940
e: bup-info@bristol.ac.uk

Details of international sales and distribution partners are available at bristoluniversitypress.co.uk

British Library Cataloguing in Publication Data
A catalogue record for this book is available from the British Library

ISBN 978-1-5292-1741-4 hardcover
ISBN 978-1-5292-1742-1 paperback
ISBN 978-1-5292-1743-8 ePub
ISBN 978-1-5292-1744-5 ePdf

Cover design: Liam Roberts
Front cover image: WLDavies

Bristol University Press uses environmentally responsible print partners.

Printed in Great Britain by CMP, Poole

Contents

List of Photographs

Timeline

List of Abbreviations

AI	artificial intelligence
CAP	Common Agricultural Policy
CCC	Committee on Climate Change
CIC	Community Interest Company
CLT	community land trust
CSA	Community Supported Agriculture
Defra	Department of Agriculture, Food and Rural Affairs
EA	Environment Agency
ELMS	Environmental Land Management Scheme
HIDB	Highlands and Islands Development Board
IAOS	Irish Agricultural Organisation Society
LSA	Land Settlement Association
NFU	National Farmers' Union
NRW	Natural Resources Wales
NTS	National Trust for Scotland
RSPB	Royal Society for the Protection of Birds
SFI	Sustainable Farming Incentive
SFT	Sustainable Food Trust
TCPA	Town and Country Planning Association
TFA	Tenant Farmers' Association

Acknowledgements

No book, however modest, is possible without the help, wisdom, support, guidance and infinite knowledge of people too numerous to mention: the experts. Not only were they generous with time and invaluable information; in the case of my family, and wife Christine, they also added one vital ingredient: patience, and support. And that applies – again – to Emily Watt, Freya Trand and all at Bristol University Press: thank you.

Although my late dad doesn't know it, regular visits at an early age to his workplace, in what I regard as Britain's finest covered market, provided early inspiration; meeting smallholders selling veg, soft fruit, eggs, butter and much else from the fields nearby began that long fascination with land and a journey into the unknown, As I moved from (the then) Cumberland to Tyneside, and hence to Manchester and, now with family, to Glasgow and Edinburgh – then back again – the fascination grew, the journey continued, as work, and leisure often took us to most corners of Britain, Ireland and, inevitably, the glorious Highlands and Islands; the unbeatable combination of mountain, loch, sea and breath-taking beaches of Scotland, not forgetting the magnificence of Donegal, West Cork and Kerry along the way.

In this venture, so many provided both guiding hand and fresh insights – not least, the late Professor Philip Lowe, founder director of Newcastle University's Centre for Rural Economy.

Thanks, then, to Julia Aglionby, Jo Lavis, George Dunn, Sally Shortall, Jason Beedell, Kate Henderson, Paul Burrows, Rob Gibson, Ruth McAreavy, Christine Morrison, Martin Collett, David Fursdon, Tim Lang, Brian Wilson, the late Sir Peter Hall, Hugh Ellis, Jeremy Phillipson, Guy Garrod, John Dunning, David Frew, Bob Reid, Andrew Francis – and many more from the smallest farm to the largest agri-business; and my good friends at Newcastle West End Foodbank (and their inspirational market garden). Finally, special thanks to the countless interviewees who proved so generous with their time and provided fresh insight and my walking friends in Wylam Rumblers.

Love, as ever, to Christine, Laura, Mairi, Andrew, Craig, Dominic, Wilbur, Hamish and Edie. Their journey across our land – and mine – continues.

Introduction: Going Local

From the doorstep we can hear sheep bleating over the garden wall. An array of beef cattle and calves are pounding nearby fields. Beside them, the barley and wheat have been ripening alongside tall batches of broad beans – doubtless heading for livestock feed – while the remnants of an insurgent called oil-seed rape, yellowing the landscape in late spring, has morphed into a sickly tangle ready for harvesting.

If it rains heavily – early 2020 was particularly wet – we can be sure that torrents of water will tumble down nearby hillsides, rolling onto road and pasture sometimes hardened by livestock, rather than being absorbed into cultivated land or tree plantations. Drainage – good land management – sometimes seems an afterthought.

Within a semi-circle of, say, ten miles around our house, many of the challenges and opportunities present on *our* land can be found in microcosm: plenty of cattle, sheep and cultivated acres. But why so many arable acres devoted to grain and crops for feeding livestock? Why so relatively few woods – and we're luckier than many in our small corner of Britain – in a country with a lower tree cover than elsewhere in Europe? Why so little attention paid to land drainage – and, hence, to addressing the climate emergency? So many questions left hanging in the air. And so little time to address what amounts to a looming crisis: feeding Britain and preparing for the impact of global heating. These twin issues, inextricably linked, must be addressed holistically.

Walking countless country miles during three COVID-19-induced lockdowns in 2020–21 provided time for reflection: across cultivated field, hardened pasture, river banks and flood plain; through deciduous woodland and magnificently restored 'hills' of industrial waste, planted with birch and rowan; along stretches of the long-distance Hadrian's Wall path barely half a mile from our front door, and, hence, gingerly into the Northumberland National Park.

As that lockdown began to bite in the spring of 2020, I threw caution to the wind in a national newspaper column: 'Go local. It's a no-brainer'.[1] Here we were, in England's northernmost county – barely a 40-minute drive to Scotland – happily by-passing big supermarkets secure in the knowledge that local shops and food networks would keep us going. They didn't let us down. The well-stocked village store, next to the resourceful greengrocer, kept us fed and watered. Ditto: a range of other shops.

In the most challenging of times – and for countless souls and families this remains a truly dreadful period, extremely raw – here was a ray of hope on our doorstep, hopefully a pointer to a better future. 'If this crisis has told us anything, it's that habits will have to change – and that the potential for shopping and trading on the doorstep, with connection to local supply chains, is enormous,' I wrote, hopefully not over-optimistically.

In truth, from our small patch, we've been 'going local' for quite some time, further sustained by local milk deliveries from a nearby dairy as well as yet more of life's essentials from a north-east institution, Ringtons, whose black-and-green vans and ever-cheerful driver-vendors deliver tea, coffee and biscuits once a month ('tea to your door and so much more'). If that isn't enough, a local food cooperative, barely three miles distant, part of a Community Supported Agriculture network, produces every conceivable vegetable – 56 varieties, on one count – for scores of its members.

Encouraged by the headline of that column – 'The logic of local makes a crisis comeback' – I began thinking more deeply about the two overriding pressures facing our land: namely, leaving the EU, framing a long-overdue, overarching strategy for land – *the* most basic resource, after all, sadly neglected by governments – and, hopefully, finally emerging energized from a devastating pandemic. These issues present unparalleled challenges and opportunities.

Growing up in a small northern city, on the edge of the Solway plain and within sight of the Lake District, the North Pennines and the Galloway hills, made me appreciate land, and its multitude of uses, from an early age. Carlisle, literally, transcended town and country, urban and rural, England and Scotland: busy livestock marts drawing thousands from much of Britain, engineering plants devoted to agriculture and much else and a local newspaper (on which I worked) allocating pages to the life and times of farming and its employees. The annual 'price review' by the (then) Ministry of Agriculture, Fisheries and Food – whose subsidy regime guaranteed crop, milk and livestock prices for farmers – was an event not to be missed.

Back then, of course, in a country emerging from wartime, we produced approaching 80% of our food, although farm mechanization meant that an agricultural workforce, once the bulwark of rural society, was depleting. But it still employed hundreds locally – and woodland, through a state-owned body called the Forestry Commission, was expanding nearby on both sides of the border, with the creation of villages to house hundreds of new workers. The emerging border forest park was just that: it transcended the Anglo-Scottish border.

In the 1950s–1960s we didn't question where food came from because – we assumed rightly – it was often just from the fields beyond our doorstep.

In the cavernous Carlisle market, a Grade Two listed Victorian marvel of red sandstone, cast and wrought iron which my Dad managed, long benches groaned with local produce. Horticulture was then a thriving business; today it barely exists. Although wartime rationing was still a recent memory – dog-eared ration books lingered on the mantelpiece – we felt secure in our basic diet of meat/fish and two veg.

Today, as I'll explore, that security is fragile. We grow just over half the food we need. We should be capable of producing so much more to reach a respectable level of self-sufficiency. We are critically ill-prepared for the climate emergency as rising sea and water levels threaten our most productive farming areas in eastern England, overwhelming some communities. Consequently, we need to consider renewing and reworking other once-fertile areas either devoted to livestock or left fallow. We haven't much time.

Alarm bells started to ring in our daily plodge around the Tyne valley, and beyond, in 2020–21. Surely the world's sixth-biggest economy – a fragile union of three nations (and a share of a fourth) with unlimited potential – should seize this post-Brexit opportunity to rework the countryside for the benefit of all? That the UK dismally fails to feed itself represents an indictment of successive governments, a dereliction of duty by ministers who I always thought had a primary duty to keep us safe and secure. Silly me!

Should life outside the EU, at the very least, provide a wake-up call? In the following chapters, the challenges and the opportunities, personal stories and achievements of people will be outlined: growers, producers, progressive landowners, community enterprises and land trusts, professionals and farmers – all working hard for carbon neutrality, and food security, with practical measures.

The smallholders, shopkeepers and food networks closest to my home – and their counterparts further afield – give me cautious optimism, underlined by a tour of Britain in 2020–21, sometimes sadly virtual, retracing old steps. But the sights and sounds from our doorstep and beyond, and the people and the businesses going about their daily lives, provided succour amid the immense challenges ahead. They invariably delivered hope, and not despair.

1

Land of Promise

For the upright will inhabit the land, and those
with integrity will remain in it ...[1]

Around the gentle, rounded Cheviot hills, evidence of once thriving, self-sustaining settlements punctuate the landscape. Steep terraces for growing crops contour the hillsides. Ridges and furrows from ploughing are etched into valleys. Large, circular mounds provide evidence of hill forts alongside outlines of timber roundhouses in faint circles. In these magnificent uplands, amid the tumbling lapwings and ascending skylarks, agriculture survived and prospered, albeit in a near-Mediterranean climate.

In a small museum near the 11th-century St Michael's church in the village of Ingram, vivid displays of another life give a sense of the effort involved in preparing the land for relatively sophisticated farming and creating the capacity for storing food for hundreds of people, maybe more.

So rich is the archaeological treasure trove in this part of the Northumberland National Park – the fertile valley of the River Breamish and the varied uplands rich in prehistory, amid later signs of Roman occupation – that five Iron Age hill forts, collectively one of the country's largest ancient monuments, are linked in a spectacular 4.5-mile upland trail. It is a testament to the earliest agriculture.

With polished axes, aspiring farmers created patches large enough to sow cereals such as wheat, oats and barley. Domestic animals, maybe sheep, cattle and pigs, were probably introduced from the Bronze Age and into the Iron Age, roaming the uplands in a setting doubtless beloved by some of today's re-wilding enthusiasts – more on them later – until the Romans subsequently cleared the remaining woodlands.[2] "But we are reaching back to a time when there was no record," laments Chris Jones, the National Park's archaeologist. "How did they farm and for how many? What were they able to grow? We can speculate; to what extent was this subsistence farming, cooperative living, or the development of markets – a capitalist approach if you like?" Although much has been discovered in several archaeological excavations, countless acres remain untouched in this multi-coloured landscape as spring submits to summer, a rich palette of green laced with bracken, rust-tinged undergrowth blending with purpling heather and silver, rocky outcrops.

Of course, we can fondly look at the past through soil-spattered spectacles and ask why we grow just over half the food we need on the nearly three-quarters of UK land devoted to agriculture, on 217,000 holdings ranging from tiny farms to big agri-businesses.[3] Do we blame the people farming it, and a government ostensibly committed to policies which protect and enhance it? Is landownership in Britain, where the old aristocracy remains a powerful force alongside the new rich, too concentrated and acting against the national interest? If inventive people in prehistoric times could produce their own food, why can't we do even a little better today? And with our land facing so many pressures – and opportunities – why is there no coordinated strategy in England across government to address them in the round, rather than Whitehall departments, as we shall see, pursuing competing objectives?

Employing almost half a million people and contributing just 0.5% to the UK economy, worth £9.6 billion, UK farming has been receiving around £3 billion annually in subsidies through the EU's outgoing Common Agricultural Policy (CAP) – a 'significant proportion' of farm income for many, which has served only to 'dampen the focus' in the drive for efficiency.[4] This is a polite way for England's Department for Environment, Food and Rural Affairs (Defra) to say that farming subsidies have outlived their usefulness; that it's time for a welcome change, post-Brexit – which is why the subsidy regime is being phased out. Enter, in its place, a new Environmental Land Management Scheme in England, or ELMS (Scotland and Wales are developing different models), which aims to pay farmers and landowners for enhancing nature and encouraging biodiversity, so called 'public money for public goods'. There could be many farming casualties along the way.

In truth, farming can be seen in a slightly more favourable light. When the wider food and drink sector, of which agriculture forms part, is thrown into the national statistical mix a further £110 billion can be added to the economic impact, reflecting the broader contribution of farming to the UK. Naysayers might ridicule it as the last great subsidized industry, when countless farms barely make a profit in spite of government subsidy. But the ultimate beneficiaries of this state largesse are the big supermarket groups, whose profits have effectively been underpinned by the CAP, and the relatively cheap food it has delivered. And all that, it must be said, at some cost to the environment: farming is responsible for 10% of greenhouse gas emissions, partly through livestock emitting methane and chemical fertilizers spewing nitrous oxide into the atmosphere.[5] The emerging ELMS regime, you might say, has its work cut out.

A new mantra in farming, 'more for less', means initially producing at least the same amount of food, with fewer 'inputs', namely fertilizers and chemicals, and better use of soils, through 'regenerative' or 'nature-friendly' practices in a UK with one of the highest proportions of farmed land in Europe – 72% of the land mass – partly because woodland, occupying just 10%, is low by Continental standards.

Looking as an outsider at this mismatch between land and production, it seems clear that our low level of self-sufficiency in food, even after such hefty subsidies, represents something approaching a market failure. Think about it: lots of farmed land, a low level of crop production for human consumption and too many acres devoted to growing livestock feed, which accounts for a third of our grain crop.[6] Surely this is an indictment of successive governments clearly happy to let Europe and the world feed us – none more so than in the late 19th – to earlier 20th centuries, when imperial Britain ruled the waves – no matter the fragile 'just-in-time' supply lines between growers, producers, importers and the supermarket.

As Tim Lang, Emeritus Professor of Food Policy at City University, London underlines, we should be growing more when blessed with so much cultivatable land, and good soil. "Are we really saying ... the sixth-richest economy in the world, 'oh, we'll get others to feed us' when we could be growing much more ourselves'?" he asks in exasperation. "It's bonkers." Why is the issue being ducked by the British state?[7]

The following chapters, then, are an attempt to reframe the debate, recalibrate policy to reflect that farming, food production and addressing the climate emergency by restoring soils, land and landscapes form a single policy area, amid 'growing evidence that our fragmented approach to land ... means we are failing to deal effectively with the conflicts and complexities of the way we use it, and its qualities'.[8]

Given all the pressures facing our land, it seems remarkable – in the context of today's light-touch government – that land was last comprehensively examined 77 years ago in a seminal White Paper published by a wartime coalition government. It noted that 'provision for the right use of land, in accordance with a considered policy, is an essential requirement of the government's programme of post-war reconstruction'.[9] It foreshadowed a raft of sweeping legislation later in the 1940s which subsequently created a planning system, 'nationalizing' the right to develop land – now being regressively deregulated in England (but not in Scotland and Wales) – a network of National Parks and the first agricultural support regime, later absorbed into the CAP, which guaranteed farmers minimum prices for their produce.

In periods during the first half of the 20th century, the debate surrounding land use, ownership and taxation and access to the countryside was often translated into radical action – sadly, followed by official indifference in

England as land became little more than a traded commodity for the rich, a safe, tax-free place to dump spare millions; farming subsidies have thus served to inflate land prices buoyed up by high rental returns from the 48% of farms wholly or partly tenanted, as Defra helpfully reminds us.[10]

As it stands, there's little policy coherence in England where an ideologically driven, development-focused and centralist planning policy, overriding local government and promoted by one Whitehall department (Ministry of Housing, Communities and Local Government) clearly conflicts with the ELMS regime presented by another (Defra) – with the former paying 'scant regard to climate change, nature recovery strategies, biodiversity, or the whole range of competing land needs'.[11] The creation of Defra in 2001 broke the link between planning, housing and the environment – while, to further complicate policy making, the growing challenges of global warming led to the creation of a Department of Energy and Climate Change in 2008 (now absorbed into a Department of Business, Energy and Industrial Strategy). That's three government departments responsible for land, the environment, farming and climate change, with no policy alignment – and that's even before a fourth, the Department for Transport, responsible for land-hungry, carbon-heavy road building, is added. Not for nothing has John Gummer (Lord Deben), a former cabinet minister responsible for the environment (before the creation of Defra) called for a new Department of Land Use to bring these departments and disparate policy areas together.[12] Meanwhile, Scotland and Wales plot a distinctly different course with joined-up land-use frameworks: the government in the former has a statutory requirement to produce and update a land-use strategy, while in the latter well-being legislation underpins the drive for a healthy and sustainable environment.

Upland woes

All that seems a world away with farmer Ross Wilson, as we plod across the Cheviot hills, near Ingram, rich in healthy grassland. At 1,000ft, overlooking the terraces on Wether Hill, with curlews pleasantly trilling above, I ask him why this land can't be cultivated today. If one answer is fairly obvious – easy access and modern logistics might be a challenge – he marvels at the ingenuity of predecessors several thousand years ago. Ross points to a herd of 100 red deer, a new addition to the family's 1,800-acre tenanted farm, and a throwback to prehistory in those ancient field systems. "These hills were all farmed, cropped, littered with cultivation – of course, it was subsistence farming then … but all the indications are that this was a well-populated, self-sufficient area – and the earth is still good." To prove the point, he then cuts a tiny square to reveal deep soil below the grass. "See how good that is … and it goes back thousands of years."

But if our distant ancestors could grow more of their own food, why in heavens can't we do better, instead of relying on others to feed us? Where, you might ask, is the strategy across the three nations of Britain to match the idealism of the 1940s, the shared vision which underlined policy making throughout much of the last century? While accepting that domestic self-government in Scotland and Wales leads inevitably to separate policy outcomes, surely common objectives – on one island, after all – demand an overarching appraisal of land use across our three nations, once underpinned by policy coherence and progressive, often shared, land-use strategies. The independent Food, Farming and Countryside Commission has suggested that England can learn from the different approaches to integrated land use in Scotland and Wales driven by a functioning planning system, rather than a deregulated one emerging in England.

If we accept that shared interests transcend territorial boundaries on the island we can still call Britain – just! – then, surely, one key objective overrides national, or territorial, loyalties to England, Scotland and Wales: namely, food and where it's grown. As Ross Wilson and his family attest, with Scotland in full view from the Cheviot hills, distribution networks, farmers, food and livestock have no national boundaries on our small island; products, and animals, are traded across a fragile Britain. In this vital area, in short, we are one – yet, remarkably, there's little sign of a resilience strategy to underpin food production across our nations, no "unified sense of urgency".[13]

In the context of Scottish and Welsh self-government, and emerging English nationalism – for in most domestic policy areas Westminster is effectively an English Parliament – it's significant that British-wide institutions, often representing smaller-scale growers and a new wave of nature-friendly farmers, are, quite literally, a growing force and operate across boundaries. If the Westminster government, which still retains powers over the economy, appears quiescent, it's refreshing that both food producers and progressive farmers running larger undertakings are taking a lead, across Britain, in forging new alliances. And in later chapters they'll show, despite the many challenges facing our land – from ill-preparedness for the climate emergency to the fragility of farming in productive, flood-prone areas – that innovative measures and collective action are driving a new, alternative movement for better farming alongside renewed meadows, hedgerows and, sometimes, the emergence of local, alternative food supply chains.

Collectively, these dedicated souls recognize – as do some farmers with substantial holdings – that the twin pressures of leaving the EU and finally emerging from a devastating COVID-19 pandemic should present unparalleled opportunities to renew our land, rework the countryside and address the climate emergency for the benefit of all. One aim, then, in succeeding chapters, is to deliver a commodity in short supply these days: *hope*! But it comes, naturally, with a caveat.

Undeniably we are facing, at the very least, a period of great uncertainty, particularly in England, as conventional agricultural subsidies are phased out and the ELMS regime is ushered in. Some farmers are ready to call it a day; self-redundancy, far from being a parody, is the new reality for them in post-Brexit rural Britain, facing the biggest upheaval in farming and land management in the 74 years since agricultural subsidies were introduced. And Ross Wilson acknowledges that, if he were older, he might be tempted to call it a day. He sees little future in raising and trading sheep.

As it is, the Wilson family partnership is certainly blessed with enthusiasm, tinged with trepidation. Like tens of thousands of other farmers Ross, his wife Rebecca and father Johnny are being weaned off the CAP, under which subsidies were paid based on the acreage farmed – in their case, to support 900 ewes, plus around 1,200 lambs in spring, 100 cattle and scores of recently introduced red deer.

To say farmers are uneasy, fearful for the future under what they see as an ill-defined ELMS regime, is an understatement. "There'll be much less state support and I'm haunted by it ... sleepless nights," volunteers Ross, who acknowledges that big changes are overdue. "We're not making a fortune up here."

The Wilson family partnership, then, will have to survive partly on payments for farming 'sustainably' – definition to be determined – and for restoring habitats and improving biodiversity; becoming, in part, de facto land managers and perhaps creating carbon 'sinks', planting trees and diversifying as best they can. Ministers are clearly anticipating a drift from the land, with some farmers finding themselves unable to make a living under the new regime: as such, from 2022 an 'exit' payment or lump sum is being offered to those who want to get out; and on some estimates at least 18,000, mainly in the livestock sector, might retire. This all seems very rushed.[14]

Like many areas of Brexit, the consequences for food and farming were side-lined by the UK government in the run-up to leaving the EU's single market and customs union as 2021 dawned, although free-market ideologues advising ministers appeared sanguine. The radical Right seem happy to let the rest of the world feed us, regardless of questionable standards of industrial farming in, say, the USA, Brazil and, particularly, Australia, where a recent trade deal is sending shock waves through the UK farming sector and seen as a template for agreements with other countries, no matter the cost to domestic agriculture. Without farming subsidies, over 60% of all farms would be running at a loss.

While others in Defra hopefully do the number crunching, we begin plodding the upland pastures of the Cheviots on a damp, late summer morning. It's far removed from the heady arguments about the future of our land – at the extremes, intensive farming versus re-wilding – and balancing the need to protect our fragile eco-systems while ramping up domestic food production. And Ross Wilson, alarmed by the potential impact on the family's business from post-Brexit trade deals, is clearly enthused by the varied wildlife on these upland acres. "We're not stocked too heavily [with sheep] – others might be – and there are so many benefits for, say the breeding waders – nesting grouse, lapwings, oyster catchers, skylarks – but trying to cost all this is well-nigh impossible."

But with the Wilson farm being part of the Northumberland National Park, broader issues emerge, underlining a central, overarching theme in the following chapters: policy coherence across land use. That means, according to Tony Juniper, chair of Natural England (of which more later), integrating food production with nature's recovery after a long period during which they were in separate policy boxes. Potentially, in short, there's money to be made from storing and trading carbon, once a price has been agreed, thus creating a new income for hill farmers.

Whose land?

There's now a general acceptance that, as citizens of the UK, our rights over land transcend narrow ownership. They constitute a social relationship – whether material, ideological or symbolic – while the distinction between 'public' and 'private' interests, particularly in the 12 National Parks, like Northumberland, has become blurred; in other words, our citizenship, our roots in communities, localities and regions, gives us all, at the very least, a shared interest, a stake, in a land from which we all draw our identity and our very existence in three nations of Britain.[15]

To appreciate the challenges and opportunities ahead, we first need to understand the make-up of our land, its use and abuse, and to begin questioning why the status quo leaves us woefully vulnerable. The ever-resourceful poet Benjamin Zephaniah has duly obliged in a brief, ingenious aerial film; it portrays the infinite variety of our small island as the camera whisks across Britain below the clouds.[16] He calls it a 100-second 'walk'; each second represents 1% of our land. Thus: houses and gardens occupy 5% (five seconds); 6% (six seconds) is devoted to natural grasslands and 7% to 'sheep-grazed moors and heath lands'. 9% is covered by peat bogs, 'which are carbon stores', and a mere 10% is under trees. He notes that of the 27% of our land devoted to crops, 'half is fed to livestock'. And then the punch line: 28% 'takes us through pastures … the biggest single use of land is mostly used for feeding and rearing sheep and cows'.

The answer to some obvious questions – why so few trees, peat bogs and so much livestock? – is left hanging in the air. Alongside Zephaniah's abiding passion for what is a low percentage of woodland by European standards – 'I need more trees please/I think more nature would be greater' – our poet raises fundamental questions in his lyrical way: when we import at least half our vegetables, and over 80% of our fruit, why is so much of our cultivated land used for growing crops to feed livestock? And why can't some of our other livestock-pounded pastures be either turned over to crops or, in the uplands, used to create more carbon 'sinks', through restored peatland and new tree planting where appropriate – thus addressing the climate emergency and global heating?

All that's before we enter the delicate territory of how an estimated six million acres of common land, a quarter of all cultivated acreage, as well as grouse moors (the latter constitutes an estimated 550,000-acres alone in England) and much else, was effectively annexed by the landed classes from the second quarter of the 18th century to the first quarter of the 19th, 'mainly by the politically dominant landowners', according to the writer and academic Raymond Williams, who records a long process of 'conquest and seizure'. That was at time when all our land was owned by just 4.5% of the population. As the writer and ultimate chronicler of land ownership, Kevin Cahill, has powerfully underlined, these Enclosures were particularly invidious because the aristocracy took from a weak monarchy 'rights which were really those of the common people ... at a time when the common people had neither representation nor power'.[17] And those 'rights' are still being exercised with a vengeance in some areas, undermining the collective will of communities and local councils to address the climate emergency by restoring peatlands on the moors degraded by grouse shooting.

Exactly why the English, unlike the Scots and the Irish, have been largely indifferent to arguments for a degree of land reform – and, it seems to me, too often disinterested in broader countryside issues generally – is a matter of conjecture. The powerful emotions aroused by that dark period in Scottish history 200 years ago – the brutal removal of people to make way for sheep during the infamous 'Highland Clearances' – and subsequently by land reform in Ireland early in the 20th century, have rarely been similarly stirred in England by the Enclosures. But in Scotland, rhetoric and welcome government action – not least the creation of a seemingly powerful land commission – can't obscure the reality that just over 400 people, from the old aristocracy to the new rich, still own half of all private land, although, thankfully, over 520,000 acres is now owned by community trusts, with a target to double local ownership in a few years.

Compare and contrast with Ireland before partition 100 years ago. To appease the Irish Home Rule movement, the British government, through the creation of an Irish Land Commission, acquired holdings in ways which even today would be unthinkable in England, where 48% of farms are rented either wholly or in part, often from the old aristocracy and indirectly from the monarchy – the Duchies of Cornwall and Lancaster and an arms-length organization, the Crown Estate – or from a national charity like the National Trust, the biggest single farming landlord.

Eight years before the creation of the (then) Irish Free State, three-quarters of farmers had acquired the farms they tenanted throughout Ireland; today almost all Irish farms, though small by English standards, are owner-occupied, underpinned by a network of cooperative enterprises. Britain, it often seems, has barely moved on. Sally Shortall, Professor of Rural Economy at Newcastle University, who was brought up on a family farm in Ireland, says she was astounded to discover both the size of the tenanted sector in England – "and I thought I knew it all" – the continued dominance of large landowners and the scale of land trading in Britain, where relatively large holdings still come on the market. They're sometimes bought by the new rich, lured by a favourably benign taxation regime, and by the mega-wealthy from overseas; a Danish billionaire, for instance, is now the largest single owner of private land in Scotland. "Britain is the outlier in Europe," observes Shortall. "It really is the anomaly. People do not sell land in Ireland – partly it's the post-colonial legacy. You hang onto your land. You keep it in the family."

But in Britain, land trading continues apace. In spite of land reform in Scotland, 'a record high of £112m' was invested in Scottish estates in 2020 – with prices driven by a rise in investors keen to exploit emerging 'green' subsidies, plus 'increased demand from residential and lifestyle buyers'.[18] In England, investors from outside agriculture have similarly become 'almost equally important in the market' as farmers seek to expand; the former, apparently, still view land 'as a safe asset in which to store their wealth'. But there's another factor at play among non-farming investors, as we shall see later: exploiting the anticipated payments – labelled 'public money for public goods' – from the emerging ELMS regime in England.[19]

If the state, then, did directly intervene early in the 20th century to transform landownership in British-ruled Ireland, why on earth, you might ask naively, couldn't it take similar action in England, Wales and Scotland? Actually, in the latter case, it did pursue far-reaching land reform – never yet to be repeated – by creating 6,000 crofts (smallholdings) 100 years ago, through compulsory purchase of land if necessary. And a UK government also made a commitment in the mid-1960s to seriously address land reform with the creation of a Highlands and Islands Development Board (HIDB) – only to dilute subsequent legislation – while a similar enterprise planned for the North Pennines in England never became fully operational. According to the former Scottish Labour MP and UK Energy Minister Brian Wilson, who lives on the Island of Lewis, the HIDB never lived up to expectations by seriously tackling land-ownership. "Rhetoric never matched reality," he added.

1.1: Farmer Ross Wilson examines soil quality on his hill farm
Unless otherwise stated, all images are copyright of the author

**1.2: Inner-Hebrides: Eigg, owned by the community,
looking towards Rhum**

1.3: Island of Lewis: the Neolithic-Bronze age Standing Stones of Callanish, in the Western Isles where swathes of the 100-mile archipelago are now under community ownership
Source: Murdo MacLeod

1.4: The earliest agriculture?
Prehistoric farming terraces in the Cheviot Hills
Source: Ross Wilson

The English question

Experts in rural policy have long speculated why England is seemingly more quiescent. One theory is that rapid urbanization in the 19th century left most of the population detached from the land. The late Professor Philip Lowe, with other rural academics, has noted that, unlike much of mainland Europe, England has no peasantry or large class of smallholders with extended community and kinship links back to the countryside. For that reason, political resistance – unlike, say, in France – to the fluctuating economic fortunes of the countryside has been relatively muted:

> The absence of a substantial peasantry and the political clout it could mobilise has contributed to a national perspective on the countryside that is strongly urban, or consumer led … the farming community has not been valued so much for itself as for the food it could produce, and then at the lowest reasonable cost. … rural policy has been a minor adjunct of agricultural policy.[20]

Is the collective mood in England changing? Are long-forgotten links with the land being restored as more people seek sanctuary in country towns and villages to escape the density of large cities, spurred by the aftermath of COVID-19? Perhaps. But England, as Sally Shortall observes, remains an outrider, with little enthusiasm for reform in a country where an estimated 36,000 people – 0.6% of the population – own half the rural land.[21] The old landed class thus remains a powerful force, determining the scale of development in the countryside while leasing hundreds of thousands of acres to tenant farmers, some of whom have few rights. And, if anything, demand for agricultural land is outstripping supply, spurred by wealthy people with no farming background seeking a safe investment 'vehicle'.

In the following pages, then, it will hopefully become clearer through personal stories of challenges, opportunities, successes and – yes – dangers ahead, that agriculture, restoring nature, addressing global heating, alongside other complementary economic, environmental and social issues, should neatly coalesce: the need to grow more affordable crops, prepare for the climate emergency and, potentially, create more jobs in the process. All of this underlines the powerful, undeniable case for more affordable rural housing – even 're-peopling' relatively remote areas. Time and again, local food producers speak with one voice: 'The jobs are here – we have the potential – but the housing is beyond the limits for all but the well-off.' And, let's remember, housing, alongside planning, is the responsibility of the deregulatory Ministry of Housing, Communities and Local Government (MHCLG) in England – not Defra, which is supposed to oversee rural affairs.

As it is, these critical policy areas – farming, addressing the climate emergency, job creation, housing – lie in separate policy 'boxes'. It's a worryingly disparate approach to managing, sustaining and enhancing that most basic resource of all – *our* land – in which we all should have a stake, morally if not legally.

For their part, the Wilsons in the Cheviot hills – tenants of Northumberland Estates, a long-established land and property enterprise with its roots in Norman England[22] – have been working with Natural England (NE), charged with protecting and enhancing the countryside, to address nature's recovery. And according to Tony Juniper, chair of NE, the country has suffered a "catastrophic loss of nature", much greater than other countries, partly because Britain was among the first nations to industrialize. That legacy is clearly visible from the Cheviot hills where, to the east, you can spot the remnants of the county's coalfield, the industries it spawned and the more recent, dreadful impact of opencast mining some of it, in the south of the county on land owned by a local aristocrat-cum-climate change sceptic. But Ross Wilson refuses to be downbeat. Unlike other upland farmers, he can count on some arable land. "You see three fields of barley?" he gestures from the hill. "We have them in rotation. That makes sense, good husbandry. We could do continuous cropping, plough fields out – crop, crop, crop – but we'd end up decimating our soil, and we're an organic farm. So we have grass for a year, to build up organic matter on the fields, and put livestock on them."

Farming, then, is a delicate balance. Perhaps, sometimes, we need to recall the past, no matter how distant, to learn how previous civilizations survived – none more so than the Romans, who built that great, 73-mile wall from the Solway to the Tyne, and farmed in the Cheviot hills northwards, clearing forests for crops – while challenging ourselves and questioning the resilience of our food systems with alternative models. That's why the Wilsons are working flat out with others to develop local supply chains to market their meat and, hopefully ,new products. In Chapter 3, we'll see the makings of a quiet revolution in local food production and distribution.

Farming in the clouds?

Many farmers, like it or not, have been a privileged lot, largely ring-fenced from the pressures facing other industries, as one major study reminded us 35 years ago: 'The truth is [they] have received more state aid from the British government, and latterly from the European Community [now the EU], than any other industry outside the state sector.'[23] That's not all; the same study underlined that agriculture has benefited in other ways, including by favourable taxation, the de-rating of farmland and buildings and exemption from most planning control – benefits, even in 1986, estimated at between £3 billion and £5 billion.

The outgoing CAP subsidies of around £3 billion annually gave most to those with the largest slices of land – often the old aristocracy, royalty and the new rich – and least to those farming smaller enterprises; in one estimate the top 1% of the largest farms received over a quarter of all subsidies. But while some have portrayed farmers as suffering from 'subsidy addiction', Tim Lang, Emeritus Professor of Food Policy at City University, London, has, more realistically, directed his fire at big supermarket groups: "surely, if anyone it is those *after* the farm-gate who benefit from being sold otherwise unprofitable produce at low prices ... subsidies are actually a transfer from the public purse to those who benefit from ... 'cheap' food supplies".[24]

But at what cost to our land, to nature and to biodiversity? What's significant about Lowe's 1986 study is its foresight in warning about the depredation being wrought on the countryside by modern farming practices.[25] This onslaught on nature, however unintended, amounted, Lowe contended, to another agricultural revolution after 1945 – in 'scale, speed and impact' surpassing even an earlier, 18th-century revolution which saw the demise of open-field farming. This new revolution delivered not only the mechanization of agriculture but also artificial fertilizers and pesticides, selective plant and animal breeding, alongside greater efficiency. As a result, yields of wheat and barley alone virtually trebled in the post-1945 years, inflicting 'tremendous pressure' on the environment. Even then, with some prescience, the study lamented the impact on soil fertility and water quality from oil-based derivatives, making the performance of the farming sector 'far less rosy ... many natural features have been obliterated, transforming both the rural scene and the natural resources of the countryside'. A Countryside Review Committee, on which the former Ministry of Agriculture was represented, had seen warning signs as long ago as 1978, noting that 'the really significant dimension' of recent developments in agriculture was that 'nearly all of them have been unfavourable to wildlife and landscape'.

We were warned, then, decades ago. In a country rightly obsessed with maintaining a high level of domestic food production, governments of right and left were content to let an unofficial alliance between the old Ministry of Agriculture and the National Farmers' Union call the shots – whatever the impact on nature. And farmers were generously supported as the country moved from food shortages to a surplus. As the historian David Kynaston has pointed out, the period after the Agriculture Act of 1947, which guaranteed farm prices – forerunner of the CAP and, hence, another generous subsidy regime – proved to be 'jackpot time' for many farmers, if not for biodiversity, landscape and nature.[26]

The academic and historian Howard Newby was blunter. He lamented the creation of large field systems, and much else, which had left 'uprooted hedgerows, ploughed-up moorland, burning stubble, pesticides, factory farming ... sometimes our idyllic image [of the countryside] stops us seeing reality'.[27]

Ironic, then, that the emphasis today, underpinned by new legislation in which the ELMS scheme forms a backbone, is aimed at righting the wrongs of the post-war years: replanting hedgerows, crucially restoring peatland, planting trees where many have been uprooted, eliminating pesticides and, in some places, from the Scottish Highlands to Sussex and beyond, 're-wilding' tens of thousands of acres.

Some large farms, such as the Holkham estate's enterprise in north Norfolk, have embraced nature-friendly farming alongside crop production; and in Cambridgeshire, one enterprising farmer on a smaller, county council-owned farm is leading the way nationally with 'agro-forestry', planting hundreds of apple trees around his arable fields to stabilize the soil and increase biodiversity on the edge of the Fens.

Mea culpa?

Five months into the COVID-19 emergency of 2020, one powerful individual at the heart of the food industry made a startling admission: we were, ventured Dave Lewis, departing chief executive of Tesco, wasting our most precious resource – land – through a mixture of short-sightedness and the government's failure to recognize the apparent absurdity of too many animals (32 million sheep and 10 million cattle) occupying too much land which should be used for crops to boost self-sufficiency, thus creating a more 'resilient, sustainable, equitable system'.[28]

Putting aside the belated hand-wringing – who knows, maybe self-criticism? – and his call for 'tough decisions about efficient land use' there is of course the inconvenient truth that supermarkets have been indirectly subsidised through the taxpayer-funded CAP – and, hence, made extremely profitable. Rather late, you might say – if, nonetheless, surprising – for the Lewis intervention. Whether it goes a little way toward chiming with a broader argument championed by, among others, Tim Lang and, perhaps more radically, by Sir Ian Boyd, former chief scientific adviser to Defra in England, is a moot point. The former, a one-time hill farmer, asserts, with some justification, that too much land has been devoted to animal production at the expense of crops for human consumption, and wants a phased reduction of livestock; the latter wants sheep removed from the hills and vacant land transformed into woodlands and natural habitats to restore wildlife and address global heating. The resulting food gap, he insists, could be filled by producing crops indoors in controlled conditions: so-called 'vertical ' or hydroponic farming.[29] If that sounds an extreme solution, it's not that far removed from the view of the Committee for Climate Change, chaired by the former Conservative Environment Secretary John

Gummer (Lord Deben), that 'deep [greenhouse gas] emissions reduction in agriculture and land cannot be met without further changes in the way UK land is used'.[30]

Here we enter delicate territory, seemingly occupied by two extremes: those arguing for fewer animals and, thus, seen as anti farming; and those defending the status quo, crudely labelled 'subsidy junkies', dependent on EU handouts and once opposed to little change – but now reluctantly accepting that the ending of direct farming subsidies in 2027–28 will, inevitably, mean a drift from the land, hopefully with a 'golden goodbye' payment from Defra, UK Treasury permitting.

In what sometimes seems a clash of ideologies, a middle course is emerging: namely, combining modest levels of livestock with more nature-friendly schemes to restore wildlife and biodiversity; in short, a system favoured by progressive livestock farmers, notably the Lakeland hill farmer, consultant and writer, James Rebanks.[31] In calling for pragmatism, he argues that the polarized view characterizing farming as 'bad' and nature as 'good' misses the point; nature and farming can co-exist. In the hills above Ullswater, Rebanks is creating a mix of over 50 varieties of grasses and wild flowers in meadows where sheep graze, encouraging skylarks and meadow pipits in the process. 'We need to bring the two clashing ideologies about farming together to make it as sustainable and biodiverse as we can,' he asserts.

But it's clear, whatever these arguments, that Brexit is heralding a revolution on the land, more intense and far reaching than the last one through the 1947 Agriculture Act, which introduced production subsidies and farming on an industrial scale, whatever the environmental cost, and the legacy of the partly degraded landscape which emerged. The question today, of course, is whether the post-war level of food self-sufficiency – delivering around 80% of our needs at one stage – could have been reached without intensification driven by high inputs of fertilizer and pesticides and the consequent assault on nature. Certainly, at the very least by the 1970–80s, the alarm bells were sufficient from a growing body of opinion that governments – drivers of change in propping up farms with hefty subsidies, after all – could and should have taken the lead in reversing practices to help renew nature. Whether this might have had a negative impact on production is open to question.

Beyond farming

As it is, rural policy in England has changed little since immediately after the Second World War against a background of rationing and food shortages. Essentially, it viewed the countryside purely through an agricultural lens, with the wider rural economy – and still less nature – an afterthought.

But arguments go deeper than this. Does this characterization of rural policy by historian and academic Howard Newby still hold sway today: 'the product of an unholy alliance between farmers and landowners who politically controlled rural England and radical middle-class reformers who formulated post-war legislation'?[32] Whatever the arguments, the rural sociologist, Professor Mark Shucksmith, holds onto the view that successors of this post-war elite, 'middle-class, home owning incomers' continue to oppose rural development to preserve their countryside idyll – which, of course, raises fresh questions about a new wave of country dwellers, driven – literally, in a car-borne society – to the countryside by the COVID-19 pandemic.

In fact, since 1971 the population of rural England has grown faster than that of urban England, with older incomers bringing with them considerable wealth. But the growth of what Shucksmith calls 'counter urbanization' – commuting from countryside to city (until, at least, COVID-19 struck with a vengeance early in 2020) – has 'uniquely in the western world' increased house prices above those in bigger towns and cities.[33] In Cornwall over the past year (2002–21), for instance, house prices have shot up by 10%, and parish councils are trying to improve housing affordability for locals by banning the sale of newly built properties to second-home owners or investors.

All this, of course, underlines an affordable rural house crisis; the poorly paid, bulwarks of the rural economy, are largely excluded from local housing markets while affordable rented accommodation is in short supply. That's partly because residual social housing has been largely sold off – a legacy of the late Margaret Thatcher's Right to Buy programme still active today – while lower-rental replacement homes in no way fill the gap. Planning reforms advocated by the Conservative government of Boris Johnson will only make matters worse. What chance, you might well ask, of, say, redundant tenant hill farmers getting alternative homes, still less the young with aspirations to work in the countryside if the housing is available? Yet, in spite of affordable housing being in short supply, it seems remarkable that the (non-farming) rural economy remains so relatively strong: a total of 537,000 businesses in England alone, employing around four million. Where, you might ask, do the service workers live? Some of them 'reverse commute' from big city to country – in the opposite direction to the new, well-heeled rural elite.[34]

Back to the future?

Sometimes we need reminding that the more recent past can provide hope for the future. In the Cheviot hills, and in the tiny village of Ingram, it's not so long ago in historical terms that agriculture played a bigger role in the Breamish valley, where cereal crops are still grown; the Wilsons' farm delivers some barley, and oats, and possibly could produce more. Records dating back to 1600 detail three corn mills – the last surviving one at Ingram itself, is shown on a map of 1769, but is now a 'tumble of stones and unrecognisable as such.'[35]

History tells us something else: that by the late 18th century many parts of the UK had a fully commercialized farming system, innovative and driven towards serving an expanding market in food for a burgeoning urban population. By the beginning of the 19th century one fifth of the population lived in cities and towns; 50 years later, the urban population had overtaken the rural; by the end of that century four-fifths of the population was urban. Cheap food became the priority. But instead of looking to domestic producers to feed this rising population, supplemented by immigration from Ireland – and here's the warning – Britain looked overseas to North America, Australasia and South America. While neighbours in Europe protected farming with tariffs, Britain remained resolutely committed to 'free trade' – ironically a mantra of the Brexiteers today, who sometimes fondly believe this imperialist order provides a model for the UK outside the EU; indeed, in leaked e-mails one Treasury adviser argued early in 2020 that the food sector is not 'critically important' to the economy and that agriculture (and fisheries) 'certainly isn't'.[36] Can these people be serious? "You bet," one expert close to Defra told me despairingly.

While these radical voices on the right are relatively muted, and their influence on Defra can be overstated, critics of a new Agriculture Act which embeds ELMS have a point when they argue that more attention is being paid to subsidising biodiversity than to producing more food – when, in reality, the new post-Brexit order can embrace farming working in harmony with nature, as James Rebanks argues.

But we face unpalatable choices. We can't wish away the mega-farms driven by technology, covering thousands of acres and, for the time being at least, the super-efficient dairies with several thousand cows serving the big supermarket chains. But we can learn from some of the big farms already changing practices: restoring hedgerows, eliminating harmful fertilizers, improving soil quality. And, at the other extreme, we certainly can encourage local food production, building on the inspiring work of smaller-scale growers and cooperatives, and – yes – accepting the inevitable: that the structure and image of our uplands will have to change dramatically as sheep disappear from the hills, more trees are planted (in a country with low tree cover) and upland carbon sinks are created from renewed peatland and more woodland. That's before we address the case for returning some of our best farmland in the Fens and elsewhere to its natural state: 're-wetting', in the jargon. All of which leads to the emergence of a new economy on much of our land: carbon trading. The opportunities, once someone works out how to price this resource, appear endless in the medium term.

But in the short term, here's the reality: we need to grow more crops to increase domestic food production. While we might not be approaching a crisis in food security to match our lamentable record in 1939–40, when we barely produced a third of our needs and depended on imports for the rest our supply lines are vulnerable. There has been little slack in the just-in-time system, under which stock levels are kept deliberately low. As leading food academic Tim Lang says, 'a country which does not feed itself is at risk from any disruption … it's politically risky'.[37]

But now, unlike the late 1930s, we have the added crisis of global heating and extreme weather threatening our most productive acres. These issues should be in sharp focus as the most far-reaching changes in rural Britain for at least 75 years – a revolution, in some ways more profound than the transformative drive for industrial farming after 1945 – confront a nervous countryside: phasing out direct production subsidies to farmers and introducing a new regime designed to enhance rural England and encourage biodiversity.

As it is, our nations are approaching a 'collision of extremes': increasing demand for food and water amid more unpredictable weather – seemingly heavier flooding around Britain every winter and the constant threat to productive farmland from storm surges and high tides. Consequently, we should be adapting to – mitigating – the impact of climate change on that most basic resource: our land.[38] After all, surely it is a primary responsibility of government to keep us safe, secure – and fed?[39]

Through uplifting personal stories of people charting a new future, places sometimes turning the tide literally, the following chapters outline arguments for reshaping and reworking a countryside which, according to Natural England's Tony Juniper, has become "among the most environmentally degraded land in the world". Alongside the case for an upland transformation, those heartening stories involve small-scale farming enterprises growing into larger businesses – indeed an expanding movement of alternative food producers – and, occasionally, villages and small towns becoming more vibrant commercial hubs.

The case, then, for a new strategic approach to land and to the rural economy in a post-Brexit world is compelling, given the challenges and opportunities ahead and, more immediately, the social repercussions from a contraction of livestock farming. This means a coalescence of environmental, economic and social areas: renewing habitats, restoring degraded land, maybe encouraging more jobs in crop production alongside new openings in forestry, and upland management to create and sustain carbon stores through renewing peatland and, hence, trying to address flooding in the valleys below and beyond, rather than simply engaging the stop-gap solution of building defences downstream to protect villages and towns.

It is five years since I raised a leading question.[40] With our island facing so many pressures – a chronic inability to feed itself, with the climate emergency threatening our most productive farmland and coastal communities – why has the UK government no active land policy to feed, water and house our nations, while protecting them from the ravages of climate change?

It's not as if we hadn't been warned. Reports from both the office of the Government's chief scientist and the National Audit Office[41] – the spending watchdog – to name but two had cautioned that much of our best land – 'an important asset in terms of national food security' – occupies flood plains and is, thus, at risk. Today, we can add dire warnings from the government's CCC[42] and a former chief scientific adviser that a radical transformation of both farming and land use is needed – less livestock, certainly, to make way for either more domestic crops or trees and moorland carbon 'sinks' – to provide greater food security and to address the climate emergency.[43]

If, in 2015, I lamented policy stasis right at the top, what of today in a country which grows just over half its food,[44] depends largely on the EU for the rest, and potentially, faces an exodus from agriculture with 60% of farms formerly propped up by the EU's outgoing CAP?

As it is, the climate emergency – rising sea and river levels, driven by increasingly extreme weather, from wetter winters to hotter summers – risks overwhelming our best farmland and threatening coastal as well as inland rural communities. Government advisers are now openly raising the prospect of abandoning vulnerable villages and towns; one village in mid-Wales, probably the first of many in Britain, is already being prepared for 'decommissioning'. Others will assuredly follow.

Amid this escalating climate emergency, where lies confidence in the future – and hope? In those Cheviot hills, Ross Wilson is both fearful about the prospect of surviving without direct farm subsidies and cautiously optimistic that their phased-in replacement – providing public money for protecting and enhancing the landscape and encouraging biodiversity – will keep the family firm afloat. "If all (subsidy) payments ended tomorrow, we'd be in dire straits – it sits in my mind," he confesses. "We'd be out of business. I want to farm here [but] if I was 65 I'd be thinking about winding down … using a lump sum to get out." He is realistic. "I'm not saying we can't do more, We could reduce sheep numbers … We are not overstocked. We can't justify all this public money. I'm all for changing it."

Many are doing just that, changing practices radically and, as a result, facing the future with confidence by embracing nature-friendly farming, with a little help from a legacy pursued by a radical Prime Minister earlier in the last century. That now seems a golden age, when the state intervened directly to persuade the workless to work the land on newly created smallholdings while at the same time getting younger aspirants onto the first rung of the farming ladder.

If we moved heaven and earth 100 years ago, surely it's not beyond the wit of women, men and governments to reclaim at least part of our land again – for farming, carbon storage, nature's recovery and recreation. All of that, certainly, in the interests of creating a stronger rural economy, particularly in England, where there's been no debate about land use, still less any action, since the mid-1940s. 'Back to the land' need no longer be an empty slogan. And it certainly wasn't in the first half of the 20th century.

2

Learning from History

The ownership of land is not merely an enjoyment,
it is a stewardship.[1]

When David Lloyd George railed against the landlord classes as Chancellor of the Exchequer in 1909, he was consumed by the plight of farmers unable to make a decent living because of onerous conditions placed on their tenure in a country where few owned the land they worked. A year before his 'Peoples' Budget', which bore down on the aristocracy with a vengeance (relatively briefly, however) he warned owners of the land that if they ceased to discharge their functions properly, 'the time will come to reconsider the conditions under which land is held in this country'.[2] While Conservatives howled with anger, the Liberal Chancellor promptly doubled their rate of inheritance tax.

This was a time of fiery rhetoric matched by radical action over land use, rarely to be repeated; a period during the first decades of the 20th century in which a raft of measures by the state would provide the means – for those without the considerable means of the ruling, landowning classes – to gain a foothold on the farming or the smallholding ladder. It followed significant reforms of crofting in the Highlands and islands of Scotland, where tens of thousands of smallholders, dependent on subsistence farming under the constant threat of eviction, were finally granted security of tenure in 1886 and, thus, given freedoms over land and its use.

At the same time, land reform in pre-partition Ireland – still more radical than anything ever undertaken since in mainland Britain – was high on the Westminster agenda. This would lead to the break-up of the big, landed estates – effectively the downfall of the Anglo-Irish aristocracy – as farms were handed over to tenants and a network of marketing cooperatives were created. And this legacy has proved enduring. From all this, we can certainly learn from history. An active state, after all – remember that? – has its uses.[3]

Remarkably, given the vicissitudes of the subsequent 112 years – the rehabilitation and partial transformation of the aristocracy from landowners to real-estate developers; the privatization of government (and local government) land, from forests to farms; the speculative excesses of trading in land by the super-rich to avoid taxation; further attempts at land reform in Scotland – part of the Lloyd George legacy still survives in England. In Scotland, almost 15,000 crofters (or smallholders) have long had security of tenure and, more recently, the right to own their holdings – although the creation of the HIDB in the mid-1960s, ostensibly to address land reform, fell short of expectations, while a subsequent North Pennines Rural Development Board in England folded before it started serious work. But the Scottish land reform story – still underpinned by a far-reaching resettlement strategy 100 years ago – has some way to run. And one reformist legacy in England too – council-owned farms – has proved not only enduring but also a base on which to build modest expansion.

What, then, is the connection between a pace-setting arable farm in Cambridgeshire and a pioneering initiative by enlightened leaders and by Lloyd George, radical Chancellor-turned-Prime Minister, a century ago? It's not a question you'd normally ask on Whitehall Farm, where hundreds of apple trees around the fields of cereals and other crops provide an unexpected relief from the flat lands of East Anglia.[4]

Beneath the big skies of the Fens, Stephen and Lynn Briggs have transformed their 250-acre tenanted farm near Peterborough into one of the country's most varied rural enterprises, an organic undertaking par excellence where fruit and vegetables, salad crops and 15 varieties of apples are grown alongside the more traditional wheat, oats and barley. Oh, and the juice pressed from these apples is very nice too!

The new world of agro-forestry, as pioneered in a modern setting by the Briggses, might clearly have struck a chord with an early 20th-century Prime Minister who was instrumental not only in creating the Forestry Commission but also, crucially, in laying the foundations for thriving enterprises like the Briggses' undertaking. As an organic farm, it ticks all the right biodiversity boxes and is accredited as one of Natural England's highest-level stewardship schemes; truly, a pacesetter, with Stephen Briggs the leading evangelist for a new countryside order, travelling endless miles to spread the message of nature-friendly farming and agro-forestry.

Aside from the enthusiasm of the Briggses, first-generation farmers with a professional background in agronomy, the unlikely landlord should take a little credit for this enterprise as owner of the rich soils near Peterborough. Step forward Cambridgeshire County Council, one of around 60 counties which still own over 200,000 acres of farmland in England and a small amount in Wales. Although the acreage has shrunk by over half since the late 1970s, these council assets trace their history back to the vision of early ministers in the late 19th and earlier 20th century, then and now – although now in smaller quantity nationally – still offering a first rung on the farming ladder, providing land and buildings for newcomers to a sector with high up-front capital costs.[5]

The origins of county farms stretch back to a late Victorian agricultural depression, driven by imports of food primarily from South America and Australasia. Memories of the enclosures, when the rural poor lost millions of acres of common land to the aristocracy, were still raw. Widespread pleas for land reform led the Liberal MP Joseph Chamberlain – best known as a pioneering mayor of Birmingham – to stand for election pledging 'three acres and a cow' for the landless with aspirations to farm. He proposed that county councils could buy land and lease it to tenant farmers. A series of Acts in 1890, 1908 and 1925 turned his vision into reality, with the help of Lloyd George. Farms were seen as an essential 'ladder of opportunity', while further legislation in 1970 underlined their importance.

In a country where private landlordism still accounts for 48% of agricultural land, either fully or partly tenanted, Cambridgeshire has the country's largest farming estate, incorporating 180 farms spread over 33,000 acres, making the authority one of the larger landowners in East Anglia. And it has been on the acquisition trail.[6]

"The council has been very supportive," acknowledges Stephen Briggs. "We've turned this farm round [in 12 years] and we couldn't have done it without them. Rather than becoming bigger as an enterprise, we've diversified to become more efficient and, of course, organic."

For Stephen, that diversification has meant re-establishing a long-lost connection with nature while using trees which act not only as a windbreak around the fields, protecting top-grade soil from erosion, but also as a means for capturing sunlight and nutrients, and encouraging what he calls "beneficial insects" such as slug-eating ground beetles. The apples also provide a valuable cash crop, spreading the risk from a volatile grain market. Alongside, acres of pollen, and nectar-rich wild flowers provide a haven for wildlife and farmland birds: hedge sparrows, yellowhammers, reed buntings, partridge, to name a few.

Like Cambridgeshire, the neighbouring county councils of Norfolk and Suffolk, with 16,800 and 12,400 acres respectively, are similarly committed to maintaining the size of their farming estates, selling some land for housing, often affordable, while buying other parcels where appropriate. The same can't be said of other counties, notably Herefordshire, which sold most of its 4,200-acre estate in 2018, amid a public outcry, to fill a hole in a council finances and raise £40 million; a central government grant had shrunk from £60 million in 2011–12 to just over £5 million in 2018–19. Austerity, in short, has taken its toll on council land holdings; 15,700 acres alone has been sold since 2010. But not in Cambridgeshire. In January 2021 it bought its first farm in almost 50 years – a 240-acre arable holding, surrounded by other county farms – setting the trend for councils elsewhere who, history might judge, have invested unwisely over the past decade in office blocks and shopping centres.[7] Farmland, after all, is a much safer proposition – "a valuable asset", insists Mark Goldsack, chairman of the commercial and investment committee of Cambridgeshire County Council.

In this land of promise, what can we learn from history and the lessons of Lloyd George? Well, neighbouring Norfolk County Council sees assets growing – both literally and figuratively – from its 16,900-acre estate alongside crops produced by 145 tenants.[8] For a start, it stresses their importance in providing a framework for local goods, sustaining and creating rural employment beyond the farm gate, improving biodiversity and providing space both for recreation and for affordable housing. And, for its part, Suffolk, with 127 farming tenants on 12,000 acres, has undertaken an all-party review which advocated continued investing in its farming estate 'where it produces a return', along with encouraging other business opportunities, such as turning underused farm buildings into offices or retail units (or both). Cambridgeshire, similarly committed to diversifying its land portfolio, has already turned some of its land into a solar farm, with animals grazing under the photovoltaic panels. So, why the continued government indifference to council-owned farmland, in the light of its successes in East Anglia and further afield? Aside from objections from the radical Right, who think the state – local or national – should have no interest in landownership, there's another problem. According to Charles Coats, who managed Gloucestershire's farming estate for 27 years, the last substantial piece of legislation aimed at reinforcing the status of county farms – that Agriculture Act of 1970 – never lived up to expectations. Charles, the ultimate authority on council farms, recalls that the legislation laid down a "general aim" for councils to buy more land for aspiring farmers and to submit plans annually for scrutiny by the former Ministry of Agriculture.[9] However, because the farms were classed as a 'discretionary' rather than a 'statutory' service, they were never given the necessary critical oversight. Thus, he laments, they became vulnerable – and so it has proved. But at least they're still here in around 60 counties in England and, to a lesser extent, in Wales – unlike the many estates of the former Land Settlement Association (LSA), a little-known but once important part of the agricultural and horticultural landscape, which finally bit the dust in the early 1980s.[10]

The local patch

Here it gets personal. As a 'baby-boomer' growing up in the 1950s, I've fond memories of a cavernous Carlisle market (circa 1887) – one of the country's last remaining covered Victorian markets – which my Dad managed for much of his working life, along with other municipal offshoots. This was truly *the* place where a small border city (motto: 'Be just and fear not') met the surrounding countryside. And, 60 and more years ago, it was also a time when the local market, rather than the yet-to-be-created big supermarket, was the local food hub. Meat, fish, cheese, butter, veg and – yes – salad crops were raised, caught, grown, nurtured and created relatively locally. It was, in short, the end of a complex local supply chain – a source of pride to the my late Dad, who seemed to know all the smallholders by name – and a concept which a new wave of smaller-scale growers and producers are trying, often successfully, to replicate. The reasoning is simple logic: namely, that the just-in-time supply chains beloved of the Big Four supermarkets are extremely fragile and susceptible to disruption, and hardly carbon efficient when you consider the food miles of HGVs delivering from field to warehouse and then to supermarket. We can, in short, learn again from very recent history.

Every weekend in Carlisle's covered market hundreds of low, wooden trestle tables were laid out, carefully numbered, and rented by local farmers – or, more likely by their wives, who brought butter, cheese, eggs, poultry and much more to the table. But that was just one part of this extensive food offer. Crates would arrive at dawn, some delivered by tractor and many bearing the stamp 'LSA'. Soon the trestles were groaning with tomatoes, lettuce, cucumbers and assorted veg. I marvelled at the variety.

But where were the crops grown? Determined to navigate this supply chain in reverse, I tracked (on my bike) the source of this veritable exotica, now known as 'salad crops', to a multi-mile ribbon of identikit, pitched-roofed houses surrounded by a few cultivated acres and, sometimes, glasshouses as well, west of Carlisle on the road to both the Lake District and the (then) fishing vessels on the Solway coast.

The LSA, inspired by Quakers (the Society of Friends), a prominent industrialist and aided by charitable funders – the Scots-American Andrew Carnegie's foundation, for instance, still active in Britain and Ireland – was created in the early 1930s as a bold experiment with the support of Lloyd George. The aim – which today would doubtless be dismissed as either impractical or unworkable – was to persuade the unemployed to uproot from the big industrial conurbations and the coalfields for a better life learning a new skill on the land: horticulture and small-scale farming.

Like the county council farms, the roots of this collective LSA endeavour probably stretch back to a Small Holdings and Allotments Act of 1908. The idea – mind-boggling today – involved the state 'acquiring land and buildings' for plots of from 1 to 50 acres which would then be rented out for modest sums.

Two decades later, Lloyd George underlined his commitment to land reform: in a parliamentary debate on a new Smallholdings and Allotments Bill in July 1926, he thundered: "I have stated it before … I am going to state it again … this is the only country in the civilised world where the vast majority of the peasantry is landless." His task was to find them land.

Across the Irish Sea, meanwhile, a more radical concept unfolded: the enlightened Unionist MP for South Dublin, Sir Horace Plunkett, had already been instrumental in creating an Irish Agricultural Organisation Society (IAOS) – today a national driving force in Irish agriculture – while the British state was busy trying to appease Irish Home Rule by forcing landlords to sell land to tenants; memories of a devastating potato famine, to which the British state was at best callously indifferent, were still raw.

In this context, it's worth recalling the shock waves reverberating among the English aristocracy as Gladstone's Liberal government pursued land reform in Ireland. After all, while outlining plans for an Irish Land Commission in the late 19th century, Gladstone had said that his intention was to make 'landlordism impossible'. This was a time of huge uncertainty on the land not only in England but also in British-run Ireland, where grievances against a distant government were understandably higher. As the 20th century dawned – particularly after the First World War – amid a depression, high unemployment and the ever-present threat of civil unrest, land prices were low, partly driven by imports of the basic food necessities from the colonies and from the Americas. Earlier, agitation by a Tenants' Rights League in Ireland had led to the sale of estates by debt-ridden landlords. Not surprisingly, the English aristocracy in the (then more powerful) House of Lords feared their lands were under threat – as, indeed, it later transpired. A combination of high taxation and battlefield slaughter among sons and heirs of estates resulted in the biggest shift in English landownership since the dissolution of the monasteries; by 1939, over half of agricultural land in England was owned by the men who farmed it. In this quiet revolution, tenants had become owner-occupiers.

Even after 1918 in England, the mood for land reform, while more muted, was nonetheless significant, particularly among many Liberals and on the left. The embryonic LSA, then, should be seen in this reformist context alongside the county council farms initiative – all delivered by active, hands-on government; the two overlapped, with some tenants from the former moving to the latter.

A brief interlude: English land reform

For some time the Society of Friends had been piloting allotment schemes for the jobless in the aftermath of war. Progress for a bigger roll-out was slow. To hasten matters, the chairman of the London Brick Company, Sir Percy Malcolm Stewart, offered to provide £25,000 if the government would give a similar amount 'to settle unemployed industrial workers on full-time landholdings away from their home areas'. Frustrated by the slow progress, Stewart bought an estate at Potton, in Bedfordshire – a county at the centre of Britain's brick industry – which effectively forced the hand of ministers.

2.1: The Anglo-Scottish border by the River Tweed; Scotland is now adopting a more radical policy over land ownership than England

2.2: Fruit from a county council farm in Cambridgeshire devoted to the relatively new concept of agro-forestry

2.3: Wallington estate, 14,000 acres encompassing 15 tenanted farms, was given to the National Trust by Sir Charles Trevelyan in 1941 because of his disdain for private land ownership

The LSA was finally approved by Parliament in 1934; for his pains, Stewart was made Commissioner for the Special Areas, then (and still today) the more depressed areas of England. Soon the newly formed LSA had the task of 'transferring' the first 800 families from Durham, Northumberland and the then Cumberland to other parts of England (although stretches of north Cumberland, west of Carlisle, were selected as one of the new LSA sites).

A characteristic LSA landscape soon emerged, recognizable even today: a small home farm, often an original farmstead, occupied by a supervisor, with buildings for the grading and packing of produce, and beyond about 40 smallholdings of around four to eight acres each. Close to the dwellings were glasshouses, pig sheds, chicken houses, followed by a patch for fruit and vegetable cultivation and, beyond that, an area designed to be ploughed and harvested together with neighbouring plots. Sometimes there was a large orchard.[11]

This, in short, amounted to varied horticulture and agriculture on a grand, collective scale, perhaps 2,000-plus smallholdings, on 26 estates around the country, from Cumberland to Kent, Bedfordshire to Essex, which clearly caught the popular mood of the 1930s after the Great Depression. Some timber-framed, asbestos-lined prefabricated houses were developed – one company even offered a 'free' prefab for every four purchased – while eventually a fast-track system of erecting giant, seven-ton mobile glasshouses was developed to cope with the changing seasons. Against this enterprise, you can understand how the LSA thus became major producers of winter veg and salad crops – as my observations at Carlisle's market attest.

The village of Fen Drayton, in Cambridgeshire – which by 1973 had 34 acres under glass – even became an international showpiece, attracting visitors from India, Algeria and Korea to see one of the most successful LSA estates. Eventually, 'tenants' became 'growers'; preferred recruits were men (why not women as well, you might ask?) in their early 20s with at least five years' agricultural or horticultural experience, which could include three years in college. Ambition, for a time, seemed limitless; young aspirants, for instance, could tap into a special horticultural grant scheme and use LSA credit facilities. Historians noted that fields and orchards 'which had previously been in poor heart', producing a low yield of corn and fruit, were soon transformed, 'becoming a hive of industry'.

In an evocative account of the rise and sad downfall of Fen Drayton – as it happens, close to the Briggses' organic enterprise near Peterborough – Pamela Dearlove charts the ambitious beginnings of a partly successful scheme to train the jobless, mainly from a depressed north-east of England. By 1939 the LSA had almost 10,000 acres with 1,000 smallholdings and an 'efficient structure' to support the families of formerly unemployed men.[12] Although wartime brought new opportunities for home-grown food production – for the country was only producing a third of its needs – by

1945 the focus has shifted; the Ministry of Agriculture, now in charge, stipulated that new applicants, renamed 'agricultural tenants', must have farming experience – 'thus moving from a pre-war social experiment ... to a collection of agricultural cooperative communities, with coordinated supply, production and marketing ... combining to form a considerable force on the English agricultural scene'.[13]

This was no exaggeration: contracts with big food retailers – Sainsburys and Marks and Spencer, for instance – were tied up. The LSA emerged as one of the country's largest suppliers of salad crops. Tomatoes, lettuce, celery, peppers, cucumber, as well as brassicas and chrysanthemums, emerged from fields and from under giant, mobile greenhouses covering up to a quarter of an acre. They were all pioneered in this country by the LSA, as well as an emerging, new phenomenon, which also allowed year-round growing: the poly-tunnel.

Blessed with undercover growing on a grand scale, all seemed to be going well for the LSA. With emerging planting machinery aiding production, it experimented with new crops, including cherry tomatoes, packaged beans, as well as lettuces. Where they led, others followed. And as some growers progressed and gained agricultural knowledge, they moved nearby onto county council-owned farms. By 1967 a government committee of inquiry reported that the organization had attracted worldwide attention as a 'unique example of cooperation'. Its chairman, a Professor M.J. Wise, noted that the LSA provided an 'interesting lesson for many countries ... in developing a prosperous peasant economy' without destroying individual incentives for production and profit.[14]

But, having led the field, competition took its toll – particularly from the Netherlands after the UK joined the EU in 1974. Five years later, the election of a Conservative government seemed to place the LSA under a new ideological spotlight; it had few friends in an administration where state sell-offs underpinned a new political order: privatization.

On 1 December 1982, LSA growers 'listened to radios or watched their television with disbelief' as the Minister of Agriculture, Peter Walker, announced without warning that all ten remaining LSA estates, and their hundreds of smallholdings, would be sold, all jobs lost, central services such as accounting withdrawn and a central marketing system wound up by the following March. What an inglorious end – not just for the LSA but also for horticulture itself. The sector now barely exists; the Netherlands, which took inspiration from the LSA, is now a world leader in glasshouse cropping. It is also the largest single food exporter to the UK.[15]

The finality for the LSA proved gut wrenching for those responsible for integrating a complex production and supply system copied by many others. The impact on the village of Fen Drayton was profound. The central farm and its outbuildings – hub of the estate – were demolished:

In the heart of the village, the busy activities of packing and grading of salads by local people, the bustle of farm and delivery vehicles and supermarket lorries, disappeared. The manager, the advisers, the accountant were gone; the flower handling facilities were gone, the pack-house and the great propagating were gone ... all replaced by high density housing...[16]

And the legacy? As Tim Lang, Emeritus Professor of Food Policy at City University, London, has underlined, our low production of foods which can easily be grown here – fruits and vegetables, for instance – has resulted in a food trade gap increasing by over 40% from 2005–18 alone. And the UK's biggest failing lies in horticulture – the core of an LSA which once produced a large slice of our salad crops.[17]

Clearly, we cannot afford to drop to the crop-production levels of 1939, when around 70% of food was imported. But we are getting closer, with our level of self-sufficiency standing at just over 50%. While – thankfully – the UK is not approaching war, in March 2020 some feared it was coming close to an economic crisis of wartime proportions induced by COVID-19, and compounded by leaving the EU, which supplies a third of our food needs.

If we can learn anything from history, it is that food security and farming – feeding the nation – is one of the key responsibilities of government. It can't – although the current one has certainly tried – blame others. The creation, and development of the LSA, county council farms and further measures to safeguard and expand crofting in the Scottish highlands and islands, came at a time of great upheaval in rural Britain. In Scotland, returning servicemen from the First World War rightly felt cheated that, after such hardship on the battlefields of France and Belgium – having lost so many comrades in the trenches – home soil offered neither home nor livelihood. Perhaps spurred by the Irish National Land League, a Highland Land League rattled the authorities; a series of 'land raids' on the mainland and in the islands laid claim to a land long denied; the state responded with radical land reform in Scotland. First, in 1919, a Scottish-only land resettlement Act could have – but didn't – set the pace for wider land reform in Britain. Then, by the mid-1960s, the government's creation of a new institution – the HIDB – was portrayed as a long-overdue attempt to finally address a Highland question which the then Secretary of State for Scotland, Willie Ross, declared had been on 'Scotland's conscience' for far too long. And all this at a time when, in Ireland – North and South – the landownership issue, if not the wider rural problem of worklessness, had been effectively resolved before partition (in 1921) by a British state either afraid or unwilling to apply similar measures in England and Wales: namely, bearing down on aristocratic ownership.

But with the 26 counties of Ireland approaching independence – and six in the north remaining outside the then Irish Free State and inside a new United Kingdom – England was to experience its own, more leisurely social revolution: the call of the country.

Back to the land

In the 1930s, rural land prices had all but collapsed, partly due to an agricultural decline induced by cheap imports from the Americas and Australasia, which first arrived in the later 19th century. Thus emerged a free-trade Britain, and an imperial food order, quite distinct – even then – from mainland European neighbours who protected their farming with tariffs. Two other factors pushed land prices further down: the break-up of landed estates, due to Lloyd George's doubling of death duties, and the rate of slaughter among the heirs to large estates in the First World War.

With prices so low, there were several consequences. First, the LSA, and county councils, could afford to buy land. Second, tenant farmers at last fulfilled a long-held dream: thousands bought their holdings. And third, working people could buy a small parcel of land and build a modest weekend home by the sea or in the shires – for this was long before the Town and Country Planning Act 1947 laid down strict rules on what could be built and where.

Soon, modest wooden shacks, converted railway coaches and more substantial structures emerged in large pockets along the south coast, on the Downs and all points around London and other conurbations. Planners coined the descriptive term 'plotlands'; what we might call 'super-shacks' soon became an interwar phenomenon not just confined to south-east England. Around the country – even close to my doorstep – you might still see remnants of these often basic, sometimes more substantial, cabins-cum-dwellings scattered around the fields. Other places were of a bigger scale altogether; Peacehaven in East Sussex, Jaywick in Essex, for instance entered the popular consciousness.

For one observer, it was as though a proportion of the population were obeying a call of nature in seeking out a plot in the country on which they could build a substantial cabin for the family. But the late geographer Sir Peter Hall cites another reason: when the plotlands phenomenon began, most families in British cities and towns were only one or two generations away from rural life. In short, they had a connection with the land which, for many in Britain today, has sadly disappeared – but probably not in Ireland, and certainly not in much of mainland Europe.[18]

If the UK could not – or, rather, would not – use land reform in Ireland as a template for similar measures in England and Wales (Scotland, at least, briefly moved towards reform in 1919), Labour governments in 1945 and

1964 did briefly embrace the issue with legislation to create the HIDB in Scotland and, briefly, a North Pennines Rural Development Board in England, although it was effectively stillborn, abolished before it got down to business. Hopes that the 1947 Town and Country Planning Act might be a precursor to the public ownership of all land were dashed as food production became the imperative in an unofficial alliance between the old Ministry of Agriculture and a National Farmers' Union influenced by landowners.

But, on the Cambridgeshire Fens – and in other counties – the legacy of very limited public landownership lives on and still provides a first step on the farming ladder for young aspirants. And, elsewhere, a new wave of community farmers and local supply chains – a small but growing force collectively – is proving a welcome force to refocus a long-lost connection with the land and driving a radical movement for change.

3

Small is Beautiful:
The New Revolutionaries

*We should be searching for policies to reconstruct rural culture to open
the land for the gainful occupation to large numbers of people.*[1]

In the rarefied world of the landowning aristocracy, the splendid Tudor
manor in west Oxfordshire might normally be the financial hub of a modest
1,000-acre estate, providing a tidy income for the lord and master. Instead,
Hardwick House, with its extensive woods and farmland, is the centre of
a quiet agrarian revolution: to provide a relatively inexpensive space for
aspiring farmers to live their dream and to hone their craft.

Beside the chalk and the flinty earth of the Chiltern hills, rolling through
woodland down to the River Thames near Whitchurch, the estate has already
been leading the way with one of the country's older, more adventurous
organic enterprises, inspiring smallholders and horticulturalists throughout
Britain, Europe and further afield. And horticulture is a sadly neglected
area since the disappearance of the LSA in the early 1980s, alongside other
producers, as imports replaced a once-vibrant vegetable and fruit sector.

But, as we shall see, an inspirational movement is gaining momentum,
creating short, field-to-fork supply chains around Britain, created by local
growers and producers. Certainly, it's not a stretch to label the people, and
the organizations involved, as the new food – and land? – revolutionaries.
Something is stirring: from the Knockfarrell organic croft in the Highlands
of Scotland, to Canalside Community Food in Warwickshire, Cae Tan
community market garden on the Gower peninsula, the diverse Plaw Hatch
farm in East Sussex and Liv and Henry's productive three acres at Down
Farm in Devon, dedicated to "more on less land" and growing 50 varieties
of veg. These are just a few of the many enterprises creating micro food
networks throughout Britain – all Down Farm's output is eaten within a
30-mile radius – and, thus, modestly reshaping our food system to, perhaps,
create a substantial, collective base on which to build a new food order.

Then, of course, there's Hardwick, arguably a pace setter. Sir Julian Rose, who inherited the estate over 40 years ago, is an unlikely revolutionary. In refined, genteel tones he talks quietly, proudly, of redefining the ethos of landownership, through what he labels a "social experiment": providing modest plots, and maybe houses, at low rents for those keen to work the land. With his son Lawrence and daughter Miriam at the helm, he's determined to counter the image of the country estate from (in his words) "elitist, top-down, making as much as you can" into one which provides an opportunity for younger people to get a foothold on the farming ladder. "Opportunities … are few and far between because land prices in England are so exorbitant," he gently rails. "Most young people can get nowhere near affording to buy."[2]

His aim is to underpin the Hardwick ethos as "a place where some housing tenants [in the estate's 22 cottages] pay top-whack rent" and the rest, living and working on the land, are subsidized, paying "socially oriented lower rents"; and the family has aspirations to build more affordable homes. Sir Julian insists people can work on his estate with knowledge and with ability, thus contributing to its future – setting a new trend, he says, in the way land is managed and where responsibility is shared "by people with great skills but not necessarily with much money".

This, in short, is cross-subsidy with a social edge. It gets to the heart of a debate, gaining some traction, to modestly repopulate our land with smaller, more labour-intensive enterprises, whether in farming, land renewal projects addressing the carbon challenge, or in crafts from hemp-oil production to knitwear, with natural dyes, made from the fleeces of local sheep. All these activities, and more, are undertaken on the Hardwick estate, which also includes a 250-acre mixed livestock and arable farm, several smallholdings, a dairy and cheese-maker, a forestry business and Ian Tolhurst's adventurous organic enterprise, a pace setter for many in Britain and in Europe. Sir Julian describes the estate, with a population of around 80, as a "small village".

Which, you might say, is fine for Hardwick, thanks to the largesse of Sir Julian and his family. But what about the rest? Time and again, you hear an anguished question from social entrepreneurs, the inspiring risk takers behind a modest rural revival in some villages and in the wide, open spaces of Britain, namely: where can the workers on modest incomes live, often key employees, essential to any community? Answer: there's often precious little affordable, rented accommodation because most of the rural social housing stock has been sold off – and replacements are invariably few and far between, despite the valiant efforts of not-for-profit housing associations and community land trusts (CLTs). The state-enforced sell-off has depleted rural social housing levels to such an extent that they represent only around 8% of total housing numbers in the English countryside, compared with 20% in urban areas – and social housing providers have warned that government planning 'reforms', strengthening the hand of developers, will undermine plans to build more.[3] "We need to get more people on the land," sighs Ian Tolhurst, Hardwick's pioneering organic farmer. "The agricultural population needs to be increased if we're to have a more resilient food system." That's not easy when the divide between the 'haves' and the 'have nots' has widened: the comfortably middle aged at one end and the assetless, with aspirations to work the land – and unable to find affordable housing – at the other. The latter do not need reminding that 'many rural economies [are] weak, even in the prettiest, chocolate-boxy parts … low wages, and casual labour, are rife'.[4]

Even if these young aspirants have access to a little cash, the prospect of either buying or renting land can be bleak: that most basic resource, beneath our feet, has become a valuable commodity, driven by tax breaks and traded at inflated prices, with few, if any, restrictions on how it is used. At Down Farm, near Winkleigh, in Devon, Liv James knows many younger people who'd like to become smallholders or farmers, but can't find the land on which to live their dream – yet, she complains, so much land around the country seems underused. Surely, says Liv, a mechanism could be introduced for someone, like the government, to intervene if land – "such a valuable commodity" – isn't being used in a productive way: an issue at the heart of land reform in Scotland? "Ours is certainly not top-grade agricultural land, but we can make it work," she adds. "Our strength is our diversity and all the different customers we have."

As it is, the population in the countryside has been growing at a faster rate than in urban England since 1971, fuelled by city-dwellers seeking a rural idyll – labelled 'counter urbanisation' by one leading rural sociologist – and by the growth of commuting. People are moving to the shires as they become older, ironically placing more pressure on health and social care services as they age. This middle-class influx has thus given the UK the distinction of being 'unique in the western world', with houses in the countryside more expensive than those in the cities – leaving people on modest incomes, vital to sustain the land and the wider countryside, unable either to get on the housing ladder or to find that affordable rented home.[5] The result, says Ian Tolhurst, is that a shortage of affordable rural housing is, in reality, more critical than the availability of land itself: "A farm wage barely covers the cost of renting a small cottage," he laments, speaking from experience. "If more housing was available, a lot more people would come [to the countryside]" – and, potentially, to Hardwick. It's a complaint echoed by others running small rural enterprises with the potential to expand; time and again, they raise the issue as a barrier to growth and, hence, to creating more jobs and, implicitly, to raising domestic food production, albeit modestly.

Hardwick, for its part, is on a journey. In a country where 48% of the farmed area is tenanted, either wholly or partly – leaving some, on shorter tenancies, with little security – and where the old aristocracy (to which we can add royalty) wields considerable power – Sir Julian's 'social experiment' is not just adventurous. It represents a radical departure from the norms of landlordism. Quietly assertive, he challenges other landowners to follow his example and "be a little bit more flexible and humanitarian … most of them make plenty of money … [but] they always want to protect and see an outside threat rather than [adopting] a very good, evolutionary, use of land … I could be making a very tidy living, but I've elected not to."

Sir Julian has been breaking the mould since the early 1970s. Partly inspired by Fritz Schumacher's influential book, *Small Is Beautiful*,[6] with a key theme of making better use of land – and for Schumacher it was, unquestionably, the greatest 'material source' – he first turned Hardwick into a thriving, mixed organic farm delivering unpasteurized milk, as well as beef, woodland pigs, chickens, sheep and arable crops. Produce from the estate, and from other local suppliers, was sold from what was then a novelty: the farm shop. It proved a big attraction.

When Sir Julian moved on – he now lives mainly in Poland, championing its estimated 1.5 million smaller-scale farmers – another organic enterprise had already taken root in Hardwick's historic, 20-acre market garden. Ian Tolhurst has emerged as an institution, a national organic champion, lecturing around Britain, Europe and beyond when he's not tending his acres with a small, full-time team who grow over 100 crops as well as running a reborn farm shop; he's name-checked by many of the grateful small-scale food producers he's helped over the years. "At the moment, we're experiencing our most rapid and, what appears to be quite secure growth in our business," he says with guarded optimism as he reflects on the challenges, and the opportunities, of 2020. "And we make a small profit."

That growth was driven by the COVID-19 pandemic in March 2020, when supermarket shelves emptied and, briefly, produce was rationed. Consumers took fright and turned to small-scale producers and farm shops. As a result, Tolhurst says, business doubled within a few days. He recalls a "frenzy" in the farm shop: "People got scared when they went to the supermarket and saw the reality." Tolhurst Organics was overwhelmed and, briefly, had to restrict opening.

It was the same around Britain. With countless small food producers taking the strain in the spring of 2020, the Landworkers' Alliance, representing small and medium-sized community farms, wrote an impassioned letter to the Chancellor of the Exchequer, Rishi Sunak, reminding the government that their 1,200 members are 'key workers' needing help and, invariably, receiving no farm subsidies – unlike the bigger players, who have done very nicely from the EU's CAP – with the very largest raking in millions annually from the taxpayer.[7] Smallholders, farming less than 12 acres, have been excluded from these farm subsidies – apparently on the pretext that dealing with lots of small claims would be a waste of public money! In truth, these subsidies have served only to boost the profits of the big supermarkets anyway; their pricing power, built on squeezing the margins of the primary producer (farmer and grower), has pushed UK food prices 'to some of the world's lowest'.[8]

Growing fast since its creation in 2012, the Landworkers' Alliance is not just a lobby group; it delivers detailed, practical advice to a new wave of growers on how to organize production and, then, develop local supply chains. With Britain dependent on imports for 53% of its vegetables, it called in 2020 for the equivalent of a wartime 'dig for victory' emergency programme, overseen by the government: 'This is time to actively recruit new farms and local agricultural workers to "get onto the land and lend a hand".'[9]

Jyoti Fernandes, one of the founders of the Alliance in 2015, recalls the move to launch the organization and become, as she puts it, a "bit of a movement". With increasing numbers of younger people either buying small parcels of land or, more likely, renting a few acres, pressure was mounting for someone to represent this new wave of smallholders. One issue, says Fernandes, quickly became apparent: while demand for a land was high from aspiring farmers, availability was limited – a "massive problem", she says – and, worse, finance remained restricted. Why? Because likely projections of food output do not provide sufficient potential collateral on which to base security for a bank loan. "It's a massive problem, amounting to a market failure," asserts Fernandes, an evangelist for the Alliance who (with husband Doi and friends Kerry and Oli) farms a 40-acre mixed smallholding in Dorset, complete with two self-built family houses served by solar and wind power and with their own water system.

The great land question

Addressing this question – making more acres available for those aspiring to farm when prices are beyond the reach of all but the wealthy – should surely be an overarching issue in government, as it was for earlier administrations in the last century. This 'shortage' of land amid plenty, coupled with the absence of government support for smallholdings and horticulture in England, is a fundamental complaint of the new revolutionaries as well as the older, radical guard, and is echoed by Jyoti Fernandes, Ian Tolhurst, Sir Julian Rose and countless others.

With Britain well below self-sufficiency in food production, why, you might ask, should underused land remain beyond the reach of the resourceful, with barely a whimper from a Westminster government seemingly indifferent to ramping up domestic production – at a time when land reformers in Scotland want the Holyrood government to frame stronger powers of state intervention if they believe land is not being used 'sustainably'. In this context, it's surely fair to ask if governments – including a Conservative–Liberal coalition 100 years ago – could create smallholdings in Scotland on compulsorily bought land (and later in England through the LSA), why on earth can't more modest powers of intervention be considered today to meet a demand for more workable land? If Defra's proud claim is that moving out of the EU's CAP subsidy regime presents a once-in-a-generation opportunity to transform both farming and the way the countryside is managed, surely it's not a stretch to use all available powers – and new ones if necessary – to ensure potential new entrants to agriculture, and horticulture, can gain a first step on the farming ladder. While owning land, either inherited or recently bought with surplus cash, offers considerable unearned benefits – including the outgoing farm subsidy regime, and siphoning off cash in 'tax-efficient' offshore trusts – it comes with few, if any responsibilities to farm and manage the countryside responsibly. There's neither a code of conduct nor an obligation to sign up to good or sustainable management; billions, literally, have been handed out – sometimes squandered – in farming subsidies with few questions asked.

As Sir Julian Rose rightly acknowledges, as things stand, "exorbitant" land prices mean opportunities for young farmers are extremely limited. "Most can get nowhere near affording to buy," he complains. "I'm trying to make it possible for these people to gain a foothold on the farming ladder … at least, to provide a way in." Why, he pleads, do others not follow suit?

As it happens, Jyoti Fernandes sits on a Defra panel advising on how the emerging ELMS regime can be implemented. Unsurprisingly, she's pushing the case for additional help to sustain smaller, local producers – such as those represented by the Alliance, and a Community Supported Agriculture (CSA) network, for instance – and develop local supply chains.

The CSA raises a question striking at the heart of a government approach: why is so much spent on subsidizing large and medium-sized farms – but not, crucially, on smaller-scale undertakings with the potential to expand? Why, for instance, exclude smallholdings below 12.3 acres (five hectares) from the ELMS regime, which will replace farm subsidies; and why not, asks the CSA, provide training and support both for new entrants into horticulture and for smallholders wanting to gain more knowledge of agriculture?[10] Sometimes, frankly, you despair of governments' inability to back smaller undertakings with the potential for growth; support, after all, can deliver jobs, more land for food and, consequently, a more vibrant countryside.

In all this, let's remember that the government, along with royalty, is no bystander; it's a substantial landowner, through – for instance – the Forestry Commission (489,814 acres), Ministry of Defence (397,098), alongside the Crown Estate and the Duchies of Cornwall and Lancaster (436,482).

3.1: Local food cooperative 'Go Local', Ovington, Northumberland: part of the Community Supported Agriculture Network

3.2: Beans thriving in a Scottish market garden

3.3: The Land Settlement Association once underpinned British horticulture – a sector which barely exists today

Source: The Museum of English Rural Life, University of Reading

3.4: Hardwick House, Oxfordshire: unlikely setting for an agrarian revolution?

Source: Lawrence Rose

In reality, only the government can build more resilience into our fragile food system to support emerging producers and, hence, help them create these alternative local supply chains.

Whether, on the broader front, it has a statutory duty to maintain food supplies, with a 'food defence' or resilience strategy, is a moot point. As others have suggested, under the Civil Contingencies Act 2004, it does have a duty to communicate with the public about the danger of shortages. And while there was precious little communication surrounding the fragility of food supplies in the run-up to leaving the EU – and beyond – alarm bells have certainly been ringing in Whitehall, with 'rising concern about the effects on civil society from food disruption'.[11] That's because around half of our vegetables, and much else, come from the EU; all this at a time when, as Ian Tolhurst, Jyoti Fernandes and countless other smaller-scale producers attest, we could be doing so much more to expand domestic food production. While ministers appear detached, senior officials in Whitehall are not.[12] The realities – the fragilities – of the 'just-in-time' delivery systems (and the long supply chains) adopted by the big supermarket groups have been explained to ministers and, hence, the possibilities of shortages due to the lack of warehousing and storage space. Whitehall was worried. But the absence of knowledge, and concern, in ministerial circles, has been palpable.[13]

You might, then, wonder why, if a small, newish organization like the Landworkers' Alliance can raise concerns, along with many others – including the National Farmers' Union (NFU) in England – governments have been so detached from the reality of our fragile food system. This warning from the Landworkers' Alliance should be a clarion call: 'it is possible to dramatically increase domestic production of many crops, but growers ... will need government support to scale up'.[14]

Thankfully, against the odds, some producers and suppliers are taking the lead. Take Hodmedod's, an adventurous company in Suffolk pioneering 'the great British bean project' with its own, model supply chain distributing all manner of pulses and grains from a network of mainly local farmers and growers. In the first week of 'lockdown' in March 2020, as supermarket shelves emptied, sales soared 1,500%. They had to close their website for a few days to "literally take stock", according to one of the three founders, Josiah Meldrum.

In the emerging world of alternative, local supply chains – field to distributor and hence to table – Hodmedod's, with only 12 staff, is a market leader. Its co-founders were involved with different food projects in East Anglia: launching farmers' markets', campaigning for councils to source local food and encouraging farmers to go organic. With networks created, they had the novel idea of growing premium products through their farming allies: pulses, grains, legumes, for instance, which can be cultivated between rotations of cereals.

In 2014 they launched a range of fava beans, grown in Britain since the Iron Age; three years later a commercial-scale crop of lentils emerged. By 2019 they became the first company to offer British-grown chickpeas for UK sale; Hodmedod's has thus emerged as a brand with attitude. The enterprise cooperates with a growing network of 20 crop breeders and with farmers – guaranteeing to buy trial harvests and supporting tests of the country's first commercial grown crops of chickpeas, lentils and chia seeds. Nick Saltmarsh, another co-founder, talks optimistically of wider opportunities, working with growers "to produce a range of foods we can grow here."

Around Britain, local growing enterprises are similarly pushing the production boundaries, creating new networks – often through cooperative ventures, with members paying an annual subscription and, hence, signing up for regular supplies of veg, salad crops and fruit. At Down Farm, near Winkleigh in mid-Devon, Liv James and Henry Allison – new entrants to farming in 2016 – manage to grow 50 varieties and supply local shops and restaurants, as well as veg boxes, within a 30-mile radius. If they've one complaint, the couple – lucky to find three acres through a family connection – find that the non-availability of land elsewhere is holding back local food production. Liv says she's regularly meeting aspiring growers who find it impossible to find land – "we were lucky, but lots of people don't have that opportunity".

At the other end of Britain, 470 miles north, Jo Hunt solved that problem 10 years ago; he bought the tenancy of a croft (smallholding) beside Dingwall, near the Cromarty Firth in the Scottish Highlands. With 44 acres and five poly-tunnels – tenancy secured for life, thanks to legislation in 1886 – Knockfarrel Produce grows all manner of organic crops (including 80 varieties of veg) for 200 weekly subscribers. Jo, who previously ran an economic consultancy, also has 50 pigs, which means he can make home-produced bacon and sausage; handily, an abattoir is just five miles away. What's surprising is that a CAP subsidy has provided only 1.5% of the croft's £110,000 annual turnover – and he's an exception among smaller-scale growers in getting something! – against a British average of around 60%; the bigger the farm, the greater the subsidy. Jo knows all his customers by name and sends them a newsletter every fortnight – "we want them to belong to our croft."

That sense of 'belonging' proved invaluable to Canalside Community Food, near Leamington Spa, when they wanted to buy 10 rented acres near the Grand Union Canal in Warwickshire in 2017. They launched a share offer and raised the necessary £110,000 from supporters – a significant achievement made easier because the enterprise, registered as a Community Benefit Society, is literally 'rooted' in the community it serves, with 190 member households. Aside from the usual vegetables, peppers, courgettes, fennel, sweetcorn, chillies, apples and pears regularly feature in their boxes.

It's a similar formula at Gower, near Swansea. Formed in 2013, the Cae Tan community smallholding has eight acres, employs five growers and sells weekly veg boxes to 126 households, mostly subscribers. But here's the difference from England: the support of the Welsh Government, through its Natural Resources Wales (NRW) agency, which actively backs 'quality jobs and a viable return for those managing and farming the land'.[15] Cae Tan, based near the more affluent western edge of Swansea, has already developed strong links with local schools, partly through a 'plate to pizza project': primary pupils create pizza bases from wheat, grown by Cae Tan, and topped with its veg. Helped with a grant from NRW, the social enterprise now hopes to extend its reach to the less-affluent east of Swansea, linking a new market garden to another community and more schools. The head grower, Tom O'Kane, who has worked on projects around Europe, insists they've made comparisons with local supermarkets to check on value per basket: "and the price has worked out the same".

Where is this all leading, I ask Ian Tolhurst in his delightful walled garden at Hardwick House. Perhaps to the emergence of a new food order driven by local supply chains? Steady on, he cautions: "There's definitely a growing movement, particularly strong, and valuable, with a high degree of resilience ... much more able to cope," he responds thoughtfully. Sure, he ventures, it was truly incredible how lots of smaller producers, similar to Tolhurst Organics, coped with demand during the dark days of 2020 – briefly putting supermarkets to shame.

But time for a reality check, perhaps? "I've been organic now for 44 years. If you asked me then where I'd expect to be now it would be that half the food in the UK would be [produced] in the way we do it. Of course, that hasn't happened – it's only three %" (but that's on over a million acres). Why? The power of the supermarkets, he responds, the multinational food corporations, underpinned by a UK culture for cheap food at any price.

Ninety miles south, I ask the same question of Rebecca Heys at Plaw Hatch Farm, now in its 40th year as a (200-acre) community-owned enterprise on the edge of Ashdown Forest, near Sharpthorne, West Sussex. As a biodynamic undertaking, they keep sheep – handy to fertilize fields – beef cattle, a few pigs and a 30-strong dairy herd, which helps to deliver cheese and yoghurt. They grow barley, wheat and oats. Naturally, they've a cluster of poly-tunnels for year-round cultivation. Best of all, says Rebecca, the business manager, they make a small profit – £150,000 annually at the last count – and can survive without a modest £30,000 subsidy from the EU's CAP, denied to most smallholders, which will soon be replaced by a yet-to-be-fully defined ELMS scheme in England. Her concerns? The price of land – "far too expensive" – and the absence of affordable housing for people who want to work on farms and in the countryside.

But Plaw Hatch has another purpose. It's a local food 'hub', serving ten other producers – and Rebecca insists they coped well in March and April 2020 when supermarket supply lines were stretched to breaking point. "We didn't run out of food, we had no empty shelves – unlike them – although huge numbers [of people] came to us. The producers we deal with react quickly ... they're much more agile to deal with peaks."

It's the same story on my own doorstep, too. A local food cooperative – Go Local, also part of the CSA network – experienced such a surge in demand from its market garden (producing eight tonnes of veg from just two acres in 2020) that it turned to another landowner in the hope of doubling size and output. Sir Hugh Blackett, who heads the Matfen estate in the Tyne valley – nine tenanted farms in his portfolio – obliged because he simply wanted to "strengthen the local community".

Do Hardwick, Hodmedod's, Plow Hatch, Go Local and the rest tell us something more profound about the seemingly unstoppable growth of local food networks in Britain – the countless small-scale farmers and horticulturalists who, collectively, represent a growing movement and produce crops of endless variety for sale in farm shops, in boxes delivered to doorsteps, or in local stores and markets?

COVID-19, after all, underlined one vulnerability – among many others – during those dark days from early March 2020, and later on in that challenging year and into 2021: the fragility of supermarket supply chains, built on just-in-time delivery systems, and the relative strength, and importance, of a little-known phenomenon which has been quietly expanding in recent years: local growers and producers, like Ian Tolhurst's organic enterprise, delivering crops directly from farm and smallholding to table. A new wave of farmers – often either organic or organic-plus (that is, biodynamic) and nature friendly – came into their own. They are now represented by a variety of organizations dedicated to creating shorter supply lines direct from field and poly-tunnel to table: the Landworkers' Alliance, of course, the Biodynamic Land Trust, Nature Friendly Farming Network, the Campaign for Real Farming, the Pasture Fed Livestock Association, Community Supported Agriculture, to name a few.

Combined, these represent a campaigning force, crystallized each year at the Oxford Real Farming Conference, or ORFC – antidote to the more formal and long-running Oxford Farming Conference, the premier event on the UK farming calendar. Significantly, the former now draws many more than the latter – three times larger, according to Ruth Tudge, one of the three founders of the ORFC along with husband Colin and Graham Harvey, former agricultural story editor of the BBC's long-running serial *The Archers* ('an every-day story of country folk'). Ruth, who booked the hall for the first modest (half-day) event in 2010 – then, in effect a 'fringe' for the established conference – recalls the main driving force, namely, access to land. "Defra thought that anything less than 10 acres was not viable ... something we dispute, of course."

Where, then, could this be leading? Purely by chance, midway through the first lockdown of 2020, the Prince of Wales piled in – landowner extraordinary through his 130,444-acre Duchy of Cornwall lands, created in 1337 to provide an income for the heir to the throne, and one of the largest farming landlords.

"Could a transformation of our food system be one of the lasting legacies of COVID-19?" he asked on BBC Radio Four's *Farming Today* programme, while venturing that rethinking the food system – driven by a growing interest in "nature-friendly farming" – could be a "silver lining" among the dark clouds of the pandemic.[16] We shall see. After all, as a major landowner with some influence – who knows? – he might have acres to spare, perhaps?

Something certainly is stirring among the emerging local supply chains driven by an army of growers, lovingly tending the smallish plots, and the poly-tunnels of countless field-to-table cooperatives and social enterprises around Britain. And Sir Julian Rose says, he's noticed a shift over the last few years, with a growing interest – literally – from those wanting to try their hand on the land. He regularly has inquiries from people asking "might you have a piece of land for us to develop a business?" He does his best to oblige, but can't understand why other big landowners don't follow his example by displaying some "flexibility and humanity" with a "more thoughtful, less protective approach". He reflects further: "That's the problem with landowning. They always want to protect and see the outside world as a threat, rather than [seeing] a very good, evolutionary use of land ... they don't seem to get it."

In another world, or course – if not quite lord of the manor – he could be making quite a tidy living? He chuckles: "I could be making a very tidy living, but I've elected not to [and] rather make an acceptable living. That's the subtle difference."

Then, of course, there's the other extreme by way of contrast: large-scale farming, driven by technology, artificial intelligence, big field systems monitored by drones and maintained by self-driving machinery. Can these two extremes coexist? In reality, there might be little alternative as we strive for greater self-sufficiency in food production – and, hopefully, encourage a more diverse range of suppliers, with essential pump-priming from the government to encourage local supply chains as part of a (so far elusive) meaningful national food strategy.[17] But, sadly, there'll be too many casualties along the way – particularly vulnerable tenant farmers, often efficient and vital to our food supplies, whose cause has been sadly neglected by successive governments. Some face a fight for survival.[18]

4

Feeding Britain

*See how the farmer waits for the precious fruit
of the earth. Be patient ...*[1]

Amid sunflowers and grasses wafting gently in the Lincolnshire breeze, the man who invented the bagless vacuum cleaner gestures around surrounding fields to highlight the latest in farming technology. On a bright summer's day, Sir James Dyson barely stops for breath: the future of farming, he enthuses, will centre on research, science and "developing new ways of doing things, [creating] new machines". As such, expertise built around consumer durables, and the new world of artificial intelligence (AI), will prove transformative in reworking the land to achieve a goal central to the country's resilience, yet side-lined by successive governments: greater self-sufficiency in food.[2]

"We should be growing our own food; we shouldn't be importing it ... terribly important," insists Dyson, warming to a theme which you might think is – but, alarmingly, is not – a top priority for government. 'We shouldn't give up ...'[3] Yet, up to now, it seems that – shamefully – governments have. We grow barely 60% of our own food.

The engineer-turned-inventor and multi-billionaire points over a hedgerow, in a short film, to 15 acres of six-metre-high glasshouses stretching as far as the eye can see: protection for the first strawberries, with other fruit to follow, bucking the seasons with a crop ready for market in November and March. Nearby, two power plants, known as anaerobic digesters – fed on maize from surrounding fields, rather than from organic waste – provide the energy for one of the country's largest farming operations.

Sir James has already turned some of his 35,000 acres into a high-tech enterprise, using the latest gadgetry to boost food production and, he insists, to encourage biodiversity; drones filming and sending the latest information on crop and ground conditions to data analysts; large, self-guided, satellite-linked tractors with multi-screen cabs more like aircraft cockpits either ploughing ultra-precise furrows or assisting machinery harvesting peas (of which Dyson is the country's largest producer) potatoes, spring barley or wheat, for instance.

This is mega-farming on a scale which has largely escaped the public eye – or, rather, eluded a population seemingly disinterested in who grows their food as long as it's cheap – and to some extent replicated in similarly big arable farms from Elveden in Suffolk (the country's largest) to another big undertaking at Holkham in north Norfolk (earliest pioneer in four-crop rotation) and a string of others in East Anglia. Industrial-scale agriculture – a contested term – has now been applied to what we fondly called dairy farming, with its enduring image of contented cows grazing fields rich in grass and clover. Not always so in the mega dairy, however; at Grosvenor Farms, near Chester – an offshoot of the Duke of Westminster's substantial property development and real estate enterprises, at home and abroad. Here, 2,500 electronically tagged cows are housed in 350-metre-long sheds when they're not wandering to a circular, highly automated, robotic milking parlour three times daily to deliver 70,000 litres every 24 hours to a milk processor and, thence, to Tesco.[4]

You might think that farming on this scale would attract criticism from purists pursuing a dream of 'rewilding' parts of Britain – perhaps taking our island back to a dreamland of pre-history, before the human species began farming. Actually, some of them are happy to see farming concentrated on an industrial scale in – say – eastern England, provided other parts of the island are devoted to a biodiverse haven where nature is allowed to flourish with minimal interference from people. Aside from them, however, there's a broader recognition that we have little alternative than farming on this scale, hopefully supplemented by smaller-scale enterprises, if we are to increase crop production – provided the end result is food for people rather than for livestock.

To be fair, in varying degrees, some of these mega farms are often devoting large tracts of land to encouraging wildlife, restoring hedgerows and woodland, working with nature to try to restore biodiversity to a countryside ravaged by post-war farming driven by pesticides and fertilizers, no matter the cost to the environment and to the landscape in creating huge field systems. According to Natural England, a government agency, this cost has been so considerable that our land is among the most environmentally degraded globally, largely because of the demands placed upon it by a country which became among the first in the world to industrialize.[5]

Today, farming is a delicate balance between technology, nature and the environment – arguably taken to extremes with the emergence of a new phenomenon, namely horticulture in giant glasshouses, of which Dyson is offering a glimpse of the future in Lincolnshire. But at Crown Point Estate, near Norwich, and Place Farm, near Bury St Edmunds, Suffolk, two giant glasshouses, owned by a company called Low Carbon Farming, aim to grow more than 10% of the UK's tomatoes, as well as cucumbers and peppers. If the scale of this £120 million combined development seems mind-boggling – just one giant glasshouse is one-and-a-half times the size of London's O2 Arena – the soil-free production appears novel.[6]

You will, in short, be hearing more of hydroponic growing, dependent only on nutrient-rich water feeding crops rising along guide wires. This represents the brave new world of 'vertical farming' – and the two sites are claimed to be a world first in carbon efficiency – funded by Greencoat Capital, one of the UK's largest green energy funds. Clearly, there's money to be made from an emerging green revolution, not only by farming indoors (an approach favoured by a former Defra chief scientific adviser) but also in gaining government grants to increase biodiversity and harnessing – and trading? – carbon sequestration in the uplands and peatlands of the UK.

All this is against a background of climate change – as we shall see – wreaking havoc on Britain's once predictable temperate climate. At the giant Elveden farm in Suffolk, farms director, Andrew Francis, says a key concern these days is finding sufficient water to sustain a farming operation on a 22,500-acre estate, which has the distinction of being the country's single largest arable enterprise. Water management, he volunteers, now takes up as much time as managing the soil: "Access to water is a huge issue in the east of England," he tells me.

But there's a wider issue, eastward in the Fens, which threatens some of Britain's best farmland – under pressure from rising sea levels and insufficient investment in flood protection. Major decisions are needed on the billions needed to protect a huge area of once extensive wetland, drained in the 17th–18th centuries. Today it represents a delicate, nature-defying balance to maintain and safeguard prime arable land often below sea level.

This, perhaps, underlines the urgency of relocating agriculture from parts of flood-prone East Anglia to safer areas, sometimes dominated by livestock – former arable acres, and new mega-glasshouses, for instance – to take up the slack if some Fenland is returned to a natural, pre-drained state of soft salt marsh and pasture. In this way, the Fens could be "more resilient to flood risk", and hence more carbon friendly, according to one specialist who advises the government. That, of course, is a debate to be had in a UK without an overarching food policy and with a government which, up to now, has been disinterested in framing one. This was clearly a revelation to Sir James Dyson as he tentatively entered farming 10 years ago, amazed to discover that Britain's self-sufficiency – we grow, at best, 60% of what we need – is so pitifully low. "We're going to use technology, robots … interpretation pictures … all the kind of things we're developing at Dyson … that can work on a farm," he enthuses on those Lincolnshire fields, as drones film his latest machinery ploughing and harvesting prime arable acres. "The parallels between our technology and our farms are greater than you might think."[7]

That the company, Beeswax Dyson, has cost Sir James hundreds of millions of pounds in state-of-the-art machinery and energy plants is only part of the farming story for the man who also brought the world fast hand-driers, hair straighteners, washing machines and – almost (until he halted the project) – electric cars and sees great synergy between Dyson research into areas such as AI and farming. This venture is serious stuff; no matter that the company – Sir James is a leading Brexiteer – has been getting a subsidy of around £3 million annually from the EU's CAP.[8] But I'm assured by one leading farming specialist whom I know – and who has spent a working life supporting smaller-scale farming enterprises – that this criticism is misplaced: "He's too easy a target. He genuinely wants to bring the innovation he brought to vacuum cleaners to farming, with technical solutions."

In fact, CAP is being steadily reduced over a seven-year period, with a 25% cut from January 2021 for the biggest farm acreages. Beeswax Dyson, and the other large farming enterprises, naturally have the financial resources to absorb the additional costs of a subsidy-free future, albeit that they might qualify for 'public money for public goods' in a new ELMS. However, for others not so fortunate, the future is bleak. As we shall see, the attrition rate in agriculture is likely to be considerable; of the nearly 90,000 farmers getting these subsidies in England alone, 40% have been making a loss – and perhaps a third, on the estimates of Defra, could be forced off the land by a fast-disappearing CAP.[9] Tenant farmers, with little financial security, are particularly vulnerable, no matter how efficient – but that, apparently, matters little to critics of farming who point to the hefty subsidies which, in various forms, have kept the industry afloat for over 70 years. And certainly the area-based payment system of the CAP – the bigger the farm, the higher the subsidy – is hard to defend when there's little or no support for smaller-scale operations. We should remember, as argued earlier, that job creation through encouraging smaller farming enterprises, and local supply chains, could help revive some areas if the UK government (and devolved administrations in Scotland and Wales) developed a food strategy that recognized their potential.

Tenants' woes

Fifty miles west of Sir James's extensive holding, Joe Stanley was tending until recently the family's 750-acre mixed livestock and arable farm, beside the National Forest in Leicestershire, with some foreboding. The enterprise faced a triple whammy: first, elimination of the CAP could wipe out their annual profit of £60,000 at a stroke; second, it's unlikely that the landlord of this tenanted farm (and, doubtless, many others) will agree to a rent reduction to compensate for this loss of income; and third, to compound the family's problems, the ravages of climate change are bearing down on the farm and its 150 pedigree English Longhorn cattle. Losses piled up after two years of "massive" crop failures driven by heavy rains, followed by droughts, says Joe now helping to run a research and educational trust, with a working farm in Leicestershire, developing sustainable food and farming strategies. The impact of global heating has been taking its toll. "This is no longer a temperate, predictable, British climate," Joe adds despairingly. "Many farmers will spend every last penny to stay on the land until they are forced off, but we are facing a catastrophe. We will not survive as we are."[10]

Global heating, of course, is also bearing down on agriculture in southern Europe's food-producing areas, on which Britain depends for further vegetables and salad crops: indeed, throughout the year, we've effectively sub-contracted production to the vast glasshouses of the Netherlands – which, unlike Britain, didn't run down horticulture – and to Spain, which, in the jargon of the times, is 'water poor'. In short, the vast agricultural and horticultural landscape of southern Spain – on which the UK has become over-dependent – is running out of water to irrigate crops, while swathes of Britain's most productive farmland are under threat from rising sea levels, compounding a looming crisis in domestic food production.

As it is, Joe Stanley, along with fellow tenant farmers, has a more immediate problem: survival. With 48% of farms partly or wholly tenanted – and, thus, an important element in our food chain – George Dunn, chief executive of their representative body, the Tenant Farmers' Association (TFA), says he's "seriously worried some will not survive". Tenants face two key problems. First, many on relatively short tenancies could be excluded from ELMS, which will partly replace CAP and pay landowners for enhancing biodiversity – encouraging wildlife, sustaining and recreating hedgerows and meadows and sequestrating carbon, for instance – because they don't own the land on which they farm. And second, they invariably haven't the capital to modernize and diversify because, as tenants, they have no land to offer as security for a bank loan in an industry which is already heavily indebted.[11] But Defra has already noted that while the EU's outgoing CAP subsidy regime 'inflated' rents to landlords – very nice for them, you might say – their final withdrawal should see a 'reversal of this impact' – and, implicitly, lower rents. But whether landlords oblige is an open question – and one that requires an answer from government, which should have influence if it cares to use it. One large landowner, the Prince of Wales' Duchy of Cornwall estate, is said by insiders to be showing "flexibility" with rents, and holding meetings with tenants to hear their concerns. So far others, including state agencies and charities, are not, according to George Dunn.[12]

Andrew Fewings knows the problems all too well. Now in his mid-50s, Andrew became a herdsman after agricultural college – "working 70-hours a week, saving as much as I could" – before he (and wife Judith, his business partner) gained the tenancy of a 64-acre county council-owned farm in Wiltshire in 1992. Five years later, husband and wife acquired the tenancy of a larger council farm, near Salisbury. By 1997 they took another step up the farming ladder and gained the tenancy of a 336-acre livestock and arable farm at Dunster, Somerset owned by the Crown Estate – a semi-state institution-turned-substantial property development enterprise, ostensibly owned by the Queen 'in right of the Crown'. Its role is to provide an income for the Royal Family – who, privately, are already substantial landowners – through the Civil List (the Prince of Wales is funded by revenue from his substantial Duchy of Cornwall estates, which own 205 square miles of land in 23 counties).

Andrew Fewings says he "lives and breathes" farming. He is particularly attached to his herd of 200 milking cows. "They are my life," he enthuses. "It's my passion to breed good cows." The husband-and-wife team have invested heavily in the farm, particularly in a high-tech 'robotic' dairy into which the herd wanders from the pastures for milking.[13]

Imagine, then, the family's surprise when the Crown Estate, seemingly inverting Mark Twain's much quoted advice ('buy land … they're not making it any more') put the farm up for sale, along with others surrounding it, in 2016. "The excuse they gave was that they could earn much more from retail parks – and where has that got them?" he shrugs. (As it happens, the Crown Estate's 2019–20 annual report recorded a revaluation loss of £552.5 million on its regional property portfolio 'reflecting the challenging retail market'. For the following (2020–21) financial year it warned things would be much worse, with profit and property valuations 'significantly down'.)[14] According to George Dunn of the TFA, the Crown Estate has now – belatedly' – largely reversed its farm sale policy on the grounds that its varied acres could be a valuable source for carbon credits to offset greenhouse gas emissions from other areas in its business. Indeed, in a variation of the Twain theme – 'keep land…' – the 2020–21 annual report acknowledges the long term financial security of farming combined with an "ever stronger future" offered by environmentally conscious agriculture and sustainable land use – a clear nod in the direction of carbon offsetting and potential income from the ELMS regime.

The Fewingses' new landlord is Sir Michael Hintze, founder of a London-based asset-management firm, who has substantial farms in his native Australia; in 2016 he expanded his agricultural interests to the UK by acquiring land in Somerset – including the Fewingses' farm.[15] To be fair, Andrew Fewings isn't complaining too much. He's now renting additional, nearby land from Hintze – part of which is being worked by his son Scott, aged 28 – taking his rented holding to 1,200 acres. He is hoping to negotiate a longer-term tenancy. But, like many other tenants, he confesses to "uncertainty in an immensely challenging time".

And here's the irony, noted earlier: at one time a British government moved to reform landownership in (then) British Ireland, before partition 100 years ago, by effectively forcing owners to hand over farms to tenants. Interestingly, to the relief of the old landed class – which still holds sway over swathes of Britain, dominating some counties – it failed to act in England, Scotland and Wales. As a result, we are left with an increasingly fragile landlord–tenant system, often built on short-term tenancies and at risk of collapse. According to the TFA's George Dunn, a sea-change is needed not only to update this system but also, crucially, to reform a national taxation regime which encourages speculative land purchases, driven, he says, by land agents working for the big national (and international) estate agencies who see easy pickings in Britain.

Prime farming land is a valuable commodity – commanding often £10,000 and more per acre – and few countries have such an open market as Britain, with so few strings attached. In these uncertain times globally, Britain is thus is a safe place to dump spare cash, driven by tax breaks, with few questions asked. Our land, in short, often seems to be a rich persons' playground – and the very wealthiest can use another route to safeguard their fortunes.[16]

Take the late Duke of Westminster, Gerald Cavendish Grosvenor, who died in 2016. As *The Times* reported,[17] he not only had a personal fortune of £616,418,184 but an estimated £8.3 billion in family trusts. This was passed on to his only son, Hugh, who inherited the title, without incurring inheritance tax; the late Duke once told me he was a strong upholder of primogeniture, an arcane system ensuring that his eldest son would inherit his estate, and not an elder daughter. And so it has transpired.[18]

According to the Tax Justice Network, a 'glaring loophole' in the tax system lets the very wealthiest families use trusts to pass on assets outside the sphere of this tax – and, of course, Grosvenor is not only a domestic and global property company; it also has extensive farming interests and runs the mega-dairy near Chester, outlined earlier.[19]

While the term 'playground for the rich' might be contested, it's clear that a life in the country, for whose with a few millions to spare, looks increasingly attractive. In mid-2020 a rural estate agency in North Yorkshire noted that farms of more than 500 acres were commanding 'premium prices' – around £10,000 an acre, for instance – from people with no farming background who were 'changing the face of the market'.[20] In fact, according to Stewart Hamilton, a senior surveyor with the agency, the pandemic itself is a factor behind the decision of the wealthy to look for a place in the country. But he stressed the importance of balance: namely, sustaining local communities to ensure that young people gained a step on the farming ladder.

When I contacted Stewart Hamilton he cited another reason for those seeking land in the low, fertile hills of the Yorkshire Wolds: the quality of the free-draining, chalky soil, which still holds moisture – unlike land further south, more prone to flooding. He had one client further south, for instance, desperate for 2,000 acres – "he would bite my hand off if I could find it". Hamilton's rationale is that increasingly unpredictable weather patterns will determine the movement of agriculture from flood-prone land to safer pastures.

But what of those caught in the midst of this land trading: tenant farmers, like Andrew Fewings or Joe Stanley, facing an uncertain future with falling incomes (from the withdrawal of the EU's CAP) and, maybe, seemingly little prospect of a reduction of rents, dictated by landlords? George Dunn, of the TFA, is a pragmatic, measured specialist in the industry, well respected across sectors – so when he complains about agents away from North Yorkshire, in big national practices, controlling the market with short-term tenancies at as high a rent as possible – "without offering proper advice to clients [potential landlords] and being hugely rewarded in the process" – he should be taken seriously. "It is madness to think we can even begin to raise productivity ... when the most important element that we use [land] is so costly," he rails. Currently, land is largely exempt from inheritance tax, and additional relief allows the sale of one farming asset to be rolled into a new farming business – thus deferring capital gains tax; the wealthy, then, can keep endlessly buying, rolling and avoiding tax. Dunn, and others, complain that this relief serves the country extremely badly and should either be abolished – thus, perhaps bringing down the price of land – or made contingent on land being used productively.[21]

As it is, successful tenant farmers attest to 'institutional investors' paying top whack for land with little interest in farming it productively – "the very wealthy deciding they want to 'hobby farm' and able to outbid others", according to Joe Stanley. But the repercussions could be serious for Britain's most fertile areas and – as we shall see – for the uplands, where the rich, and often the powerful, are eyeing tracts of tenanted land to exploit the likely subsidies on offer from the ELMS. Carbon trading, in what some are already calling 'green gold', could thus soon represent the new upland economy.

Survival of the fittest?

Today, big farming enterprises, owned either by the new rich or by the old landed class – to be fair, some running progressive farming operations, committed to enhancing biodiversity – can afford to survive by investing heavily in new enterprises, such as soft fruit (as at Beeswax Dyson). But – perhaps inevitably – in Joe Stanley's mind the reality of farming outside the EU points in one direction: the continued rise of the mega farm and the development of super-efficient mega-dairies with herds of several thousand cattle, spurning the traditional green pastures. The Stanley family moved out

4.1: Grain ripening: thanks to its temperate climate, Britain has some of the highest crop yields in the world

4.2: All is safely gathered in: harvesting the crop

4.3: Oilseed rape, yellowing the landscape before ripening to brown and used for vegetable oil, biodiesel and livestock feed

4.4: Hard at work on the country's largest arable farm – Elveden, Suffolk
Source: Elveden Farms Ltd

4.5: Know your onions. Mechanised farming is essential to feeding the nation; Elveden, also a site of Special Scientific Interest, manages this in harmony with nature
Source: Elveden Farms Ltd

of a 150-herd dairying operation in 2005, but Andrew and Judith Fewings are hanging on.

Others are luckier. Richard Williamson, former managing director of Beeswax Dyson, thinks departure from the CAP will prove to be an exciting time; the company will invest to improve efficiency – perhaps, growing more veg and fruit, like strawberries, and certainly by investing in technology to drive down costs.[22] As Sir James Dyson has said, his family is in the farming business for the long term: "We don't speculate in farmland; we invest in it." Thus, it can address the future and, hence, diversify in a world without CAP – but with yet-to-be determined subsidies for sustaining and enhancing nature.

On a fine autumn morning, amid restored hedgerows and diverse field strips and woodland – testament to a commitment to biodiversity, insists Dyson – our conversation gets off to a difficult start when I describe Beeswax Dyson to Richard Williamson, now senior managing director of a new agri-tech company devoted to smarter farming, as "industrial" in its scale. "An absolutely horrid term, something I abhor," responds the agreeable Williamson. "I hate it." But why? "Because it is thrown around and it is not easy to define what it really means. It's used widely, emotively … in a derogatory way."

We settle on my suggestion – "using technology to get the best out of the land". Richard says that's an interesting point because, historically, technology on farms was essentially about fertilizer, pesticides, increasing production, and thus became synonymous with high-input, high-output farming. And that's certainly not what Beeswax Dyson represents, because, essentially, such 'mono-cultural farming' has proved unworkable and bad for the soil, he insists. The company therefore has reached a point where "we need to do things differently … using technology to try and produce the same or more for less [inputs]".

James Dyson, for his part, stresses that the status of his empire as a private company gives the farming business long-term security across generations: 'we do the research and development, put in huge amounts of money, and it has a long-term payback'.[23]

It's fair to say that others, with a long record in large-scale farming, sometimes observe the Dyson operation with some bemusement, a degree of scepticism and with envy – not least at the new-found enthusiasm for growing strawberries, and other fruits, under glass. As outlined earlier, what we call horticulture has been sadly run down (unlike its equivalent in the Netherlands) with the collapse of the LSA. By value, we now grow only 57% of the vegetables we consume, and a miserable 12% of the fruit. The number of apple growers alone had fallen from 3,000 in the 1950s to barely 800 by the mid-1990s, and has clearly dropped further since.[24] The once-solid Scottish raspberry industry, mainly in Perthshire and Angus, has also declined in recent years. It's a trend that an adventurous new wave of producers, such as Stephen Briggs, are embracing with pioneering projects, such as agro-forestry; in his case, on 250-acres at Whitehall Farm, near Peterborough, where 15 varieties of apples are grown on hundreds of trees planted around fields devoted to cereals and vegetables. The system allows Stephen, a soil scientist, to extend the growing season with the harvesting of cash crops moving seamlessly from wheat to apples as the year progresses. Wheat is sold to a local windmill and, hence, goes to locally produced bread; oats go for breakfast cereals; some apples are pressed to produce juice sold in the family farm shop, with the rest despatched to wholesalers. There's a further bonus for this innovative farm, too: it is now much richer in flora and fauna as part of Natural England's higher-level stewardship scheme. As such, it's a pace setter in nature-friendly farming.[25]

By contrast Elveden, 50 miles east, in Suffolk – seat of the Earl of Iveagh – is not only the country's largest arable farm. The estate of 22,000 acres, half of which is farmed, is also a Site of Special Scientific Interest, attracting a range of – sometimes rare – ground-nesting birds, such as the stone curlew. On the one hand, Andrew Francis, the farm director, says they need "data and science" to maintain nutrient-rich soils producing potatoes, onions, carrots, parsnips as well as a range of barley, wheat and rye. On the other hand, he says, the years since 2005 have been transformative. "The agricultural space looks very different. There's been a big move in landscape management, restoring habitats, and working with neighbours to enhance nature." It's a theme championed by the fourth earl, Edward Guinness, who counts himself as Irish first and Anglo second; he also has an interest in 700 acres of farmland in County Meath. He is clearly passionate about enhancing the environment, encouraging biodiversity and – most of all – maintaining the quality of the soil. He goes into considerable detail. "Soil supports life – absolutely critical to carbon lock-up." Did I know that each gram supports up to one million bacteria? He insists they have a duty to protect future generations as "custodians of a precious environment".

It was a question foremost in the mind of a genial Thomas Coke, eighth earl of Leicester, when he appointed Jake Fiennes as general manager of conservation on the 26,000-acre Holkham estate in 2018. Fiennes, previously manager of another 5,500-acre Norfolk estate, shares the passion of James Rebanks,[26] the Cumbrian hill farmer and writer, that responsible farming and biodiversity go hand in glove, and that reinforcing the latter should strengthen, rather than weaken, the former – the essence of "sustainable, regenerative agriculture", insists Fiennes.

Thomas Coke (pronounced 'Cook'), describes Jake Fiennes as a pioneer. He's keen for his estate – which operates its own farming company – to be a leader in developing, and proving, how conservation-focused farming can push up productivity while, at the same time, enhancing wildlife habitats. For his part, Fiennes jokes that he's labelled the 'disrupter' for challenging once-accepted practices. We chat after another dawn start on his daily rounds; he's just counted an estimated 26,000 pink-footed geese, down for the winter from Greenland, on an estate which embraces the magnificent Holkham beach on the north Norfolk coast. Hedgerows are being renewed and recreated, hay meadows encouraged – cornflowers, marigolds and multiple poppy varieties are all making a comeback – and wildlife is, correspondingly, increasing, with hundreds of finches returning, alongside the grey partridge, which had disappeared some time ago.[27]

A new consensus?

To argue that a common theme is broadly emerging across agriculture, between those running the mega farms, a broad swath owning or renting – say – average-sized holdings of 500 acres and the smaller-scale producers, is perhaps a stretch. But whatever the size of the farm, it's clear that the dividing line between what once seemed to be two opposing forces is becoming blurred; that restoring and enhancing nature can coexist with regenerative, low-input farming dedicated to improving soil quality.

But there's a wider divide in the uplands – between the purists, dedicated to extensive rewilding and fiercely opposed to the scale of (mainly) sheep farming, and the pragmatists, like James Rebanks, who argue that livestock farming can work in harmony with nature.

But where stands the government? In spite of assurances from ministers about the importance of food production as a new Agriculture Act received Royal assent in early November 2020, there's little sign that the new legislation provides either the will or the means to seriously ramp up domestic food supplies. It's here where reality clashes with fine ministerial sentiments. Around the time that the new Agriculture Act – first in 70 years – appeared on the statute book, I happened to be speaking to Joe Stanley, among others, about the government's commitment to food self-sufficiency and its implication for continued imports to fill the 40–50% gap between production and demand (exactly from where, post-Brexit, the government seems unsure, because a fair chunk of these imports come from the EU!).

Joe Stanley is angry. Where's this elusive food policy, he rages? The family farm had done its best to diversify – to the extent of ending a 150-herd dairy operation in 2005 because "we couldn't make it work" with the prices offered by supermarkets. Recalling that the removal of CAP, following our exit from the EU, was a cornerstone of the Brexiteers, he thunders: "Where's the plan for agriculture? It's quite clear not a single one of them … had a strategy … a vision for farming. The [English] government doesn't have one."

Shortly afterwards an (English) Farming Minister, Victoria Prentis, appears on the early morning BBC Radio Four programme *Farming*

Today and blithely addresses what we might politely call the country's food challenge – namely, that lamentable level of self-sufficiency – with seemingly little commitment to narrowing the trade deficit in food by increasing domestic production, which should, of course, be a key priority of government.[28]

Where does food production fit into the government's plans, asks the interviewer? Her reply is revealing. "Food *security* [my italics] is at the centre of the legislation; it means different things to different people, if we're honest. Having enough food as a nation is one measure; producing a percentage of what we eat is another. Policies will enable farmers to be more productive with smarter ways of picking food, producing food, going forward …"

And there you have it! Alarmingly, the government still seems prepared to let the rest of the world meet almost half our food needs – relying on not much more than a wing and a prayer. Depending on how we calculate production – by value or by quantity – we produce just over half of our food, with a particularly alarming decline in horticulture (vegetable and fruit production), following the collapse of the LSA in the early 1980s, and the absence of any subsequent government incentives to encourage the growth of this vital horticultural sector, which now produces just 18% of our fruit.

Smaller-scale producers, with their local supply chains, now employ several thousand and are, literally, a growing force, proud to be called 'labour intensive'. Beeswax Dyson, on the other hand, might be keen to embrace robotics, AI and vision technology – "all the kind of interpretation things we're developing at Dyson", says the boss – but it does employ 170, including data analysts, agronomists, even drone pilots and three specialists with PhDs – 'bright young people working for you, thinking intelligently, willing to experiment and take risks … you can improve dramatically all the time'.[29]

Much as some want, we can't 'wish away' the mega farms; they're an essential part of our food supply chain, like it or not. But smaller-scale producers, often operating with little or no subsidies – unlike the mega farms, up to now – are important too, and capable of employing more. On his 250-acre arable and livestock farm at the foot of the Quantock hills in Somerset – not far from the Fewingses' tenanted farm – Fred Price would seem to tick all the right boxes for James Dyson and his bright young people. He, too, is both bright and young. What's more, he's that rare commodity in farming: an optimist. Aged 32, ambitious, with a geography degree from Oxford, he's already breaking the mould with what he calls restorative farming. Like James Rebanks in the Lake District, and Jake Fiennes in Norfolk, Fred Price sees great merit in 'regenerative agriculture'. He thinks we've been locked into industrial-scale farming "biologically and financially", for far too long. Finding that he was using more nitrogen fertilizer, fungicide and herbicides to get the same yield on Gothelney farm – which he runs with parents Richard and Victoria – he embraced consumer-friendly farming, with the help of an apprentice employed annually from a local further education college. He calls it "reclaiming sovereignty" in a small food economy with direct links to local retailers, restaurants and a string of bakeries, from Cornwall to Bristol, who get his grain (wheat, barley, oats), and his pork from a 300-strong pig herd. They graze herbal leys (a seed mixture of grasses, legumes, clover and herbs) and cover crops on his arable acres over winter as part of a rotation system.[30] His rationale is persuasive: big industrial farms, says Fred, produce commodities in abundance, but don't control the price – and certainly have little or no relationship with the consumer. As James Rebanks attests, the land can still be worked – and his upland farm lies 290 miles away, above Ullswater in Lakeland – with healthy soil, rivers, wetlands, woodland, scrub and "fields full of wild flowers and grasses, swarming with insects, butterflies and birds". Of course, Rebanks is luckier than other hill farmers, often tenants, who have to survive with little or no outside income.[31] Others, of course, diversify.

That's certainly the case at Abbey Home Farm, near Cirencester in Cotswold country, where Hilary and Will Chester-Master have transformed 1,600 acres into what has been labelled one of the outstanding success stories in the new-age farming scene. They took over the dairy, beef, sheep and arable holding in 1991 but, rather than reduce the range of activities, they introduced hens, pigs, soft fruit, created a 25-acre vegetable garden and produce yoghurt and cheese in a micro dairy – and, naturally, they've a farm shop too.[32] But that's not all. This is an open farm, offering, as Hilary says, "educational experiences to young people who have never been in a wood or a field ... and also to anyone who can pay to come and enjoy time on the farm to help fund the latter".

Will says they've a passion for helping young people, including those excluded from school. Before COVID-19 struck in early 2020, they were offering what he calls three- to five-day "transformative experiences" for youngsters. "We're fortunate to have a big piece of the countryside and that comes with a lot of responsibility," explains Will.

I ask Will what he thinks of the mega farms, such as Sir James Dyson's undertaking. He is relatively non-committal. "It's not what we do – far removed from our reality," he shrugs, before voicing a common concern of the many smaller-scale, alternative farming enterprises. Rather than simply offering larger and conventional farming enterprises, as well as landowners – but not tenants – cash inducements to encourage natural habitats under the ELMS, he thinks it's about time the government began helping smaller-scale and alternative farmers denied meaningful support under the outgoing CAP. "We need much more diverse farming and it could easily be scaled up if the will is there."

And that's the point, insists Minette Batters, President of the National Farmers' Union (in England and Wales). At present, she says, the will might be there, but the means seems elusive. She is speaking on 22 August 2020 – symbolically, the day after Britain would notionally run out of food, having reached the 60% level of self-sufficiency.[33] The BBC's *On Your Farm* programme obligingly tells us that we grow 18% of the fruit we eat, and just over half the vegetables consumed. Minette, who runs a farm in Wiltshire, is a refreshing break from the stale and male NFU leaders of past: pragmatic, thoughtful and, where necessary, combative. So when the interviewer asks what's the point of growing tomatoes in the UK – with all the additional energy costs – when Spain, with its warmer climate, is far more attuned to fruit production, she's quick to respond:

'Well, actually, no – not at all. Spain is one of the [highest] water-scarce countries in the world – indeed, much of our fruit and veg comes from very water-scarce countries where there is a real challenge … we have the climate here that is the jewel in the world's crown for food production. It is immoral and irresponsible to put more pressure on parts of the world that do not have the luxuries we have here.'

Yet, since the binary 'in or out' question in the 2016 EU referendum produced a relatively narrow margin for Brexit – and, let's remember, leading Brexiteers once assured us that didn't necessarily mean leaving the single market or the customs union! – there's been little clarity from the government about how British agriculture is meant to operate outside the block of 27 member states. The UK government has provided little policy detail on how it plans to balance its environmental plans (in England), central to the new Agriculture Act, with the need to produce more food – still less deliver an overarching food policy built on resilience. As it stands, as the brave new post-Brexit world bites deep into often meagre farm incomes, perhaps a third of farmers might be forced off the land – and, as we shall see, few are more vulnerable than those struggling to make a living on the uplands and hills of Britain.

The Hills were Alive

A flock of sheep that leisurely pass by / One after one the sound
of rain and bees / Murmuring, the fall of rivers, winds and seas /
Smooth fields, white sheets of water and pure sky.[1]

In a magnificent sweep of uplands, unequalled in England, the Lake District meets the Yorkshire Dales along the winding Lune Gorge on a far-northern stretch of the M6 in Cumbria: two National Parks, joined at the hip, with England's highest peaks to the west and its finest natural limestone upland 'pavements' eastwards.

The six-lane highway meanders between steep, brooding hills scattered with sheep scrambling over fells and grazing on the valley floor, before rising and then rolling down northwards to the rich pastures of the Eden valley and, thence, to Scotland.

For some hill farmers, the arrival of the M6 in 1970 – and the disruption of the preceding construction work – would have been the ultimate threat to a way of life stretching back generations. For John Dunning, schooled in agriculture, and his resourceful wife, Barbara, it became an opportunity beyond their wildest dreams, although not without financial challenges and risks along the way.

Their story is a case study of how the economy of a depressed, forgotten corner of rural England, dominated by hill farming – an industry, John correctly predicted, with limited prospects – has been transformed, employing hundreds and creating local food supply chains to serve a seemingly unglamorous new venture: motorway services, which later accommodated farm shops selling local produce.

John and Barbara Dunning might seem the stuff of legend; founders of an enterprise which eventually grew from one Westmorland service area on the north-bound carriageway of the M6 to another on the south-bound one, others on the M5, near Gloucester, on the M74, beside Lockerbie –as well as a large visitor centre on the edge of the Lakeland, on the A66 near the pleasant town of Penrith.

Quietly spoken, thoughtful and considered, John is not one for sounding off. At the couple's home between the villages of Tebay and Orton, a few miles east of the M6, he quietly insists that back in the 1950s he recognized that hill farming rested on a knife edge and could never survive as a stand-alone enterprise unless it plugged into the wider rural economy. That meant diversifying to exploit the potential of a new, tourism-centred economy in the great outdoors of the Lakes and the Dales, with sheep-rearing a side line. For John, a post-war farming revolution, underpinned by the Agriculture Act of 1947 heralding wholesale mechanization, was never going to bring the same benefits to the hills as the lowland arable areas and, consequently, "unless you diversified you would not survive".

John, aged 87 – still with the family's 900-acre upland farm, run by a manager – insists their initiative had much wider implications: namely, pointing the way for others to follow in moving the uplands away from a dependence on hill farming and sheep. Sixty years on, it may be too late for many. The hills soon won't be alive with sheep.[2]

That over 12,000 hill farms are still surviving in England alone – albeit barely and, for some, not for much longer – is as much a testament to generous subsidies from the EU's CAP as it is to individual endeavour. Those subsidies will soon disappear in England (but, significantly, not in Scotland immediately) – as John Dunning cautioned long ago – to be replaced by the ELMS, which will pay landowners and farmers for restoring natural habitats, such as peatlands and hedgerows, and planting more trees. But it's not clear where agriculture – either growing crops or raising livestock – figures in the complexities of this yet-to-be-finalized regime.

Opportunity knocks

The extension of the M6 through the old counties of Westmorland and Cumberland (since combined as 'Cumbria') thus came as a heaven-sent opportunity to the Dunnings. John recalls that the motorway shaved off one corner of his farm, where the land rises from the Lune Gorge to the summit of Shap fell, over which the old A6 passes.

What to do? Salvation came when the chief engineer responsible for designing and building the motorway told him they wanted an upland service area no matter the sparse population. There was a problem: the major operators running services on the rest of the motorway network thought such a venture couldn't be profitable in a sparsely populated area. Might John and Barbara be interested? They certainly would.[3]

We meet in the study of the couple's attractive farmhouse, a few miles from junction 38 of the M6 where the family also run a 'Truckstop' service area for goods drivers and their vehicles. Surrounded by his beloved Westmorland fells, John presents me with a copy of a recently completed autobiography, which takes up the story: 'As I listened [to the chief engineer] a wild thought crossed my mind. It resonated with my view of the need to widen the upland economy; we had reached the limits of our farm. I discussed it with Barbara: "should we give it a go"?'[4]

They did. Having no commercial catering experience, and needing capital, the couple sought partners from a local bakery in nearby Penrith. Westmorland Motorway Services was born. Still the only independent operator on the British motorway network, it opened for business in the spring of 1972 and has never looked back; a 53-room hotel followed, beside the north-bound services; a separate south-bound area subsequently opened, along with a large visitor centre for the Lake District National Park further north, Rheged, near Penrith. Farm shops, serving only local produce, were added in 2003. By then, 550 people were employed in the enterprise. Many more in the local supply chain became dependent on it.

Yet John Dunning remains uneasy. He believes the hills, and the uplands, of his native county – and elsewhere – missed a golden opportunity to develop beyond farming into tourism, and other enterprises, way back in the 1960s when a Labour government had grand plans to reform land use in his patch with a North Pennines Rural Development Board. Although legislation was approved in 1969, the board was abolished by a Conservative government, elected in 1970, before it had time to function. By then a similar organization in Scotland, the HIDB – on which the North Pennines Board would be modelled – was up and running, and it is still operating in another guise today as Highlands and Islands Enterprise. The then Labour government planned a similar venture in Wales, although it encountered strong opposition.

Records of the North Pennines initiative, far sighted and certainly with relevance today, make fascinating reading: a 3,000-square-mile area, one fifth classed as common land, stretching from the Scottish border to Skipton, in North Yorkshire.[5] While incorporating the Northumberland and Yorkshire Dales National Parks, even then two-thirds of the 6,000 farms were considered too small to provide a decent living. The consequent aim was to encourage farm consolidation, with limited land-acquisition powers, while providing grants to diversity the rural economy away from farming and towards tourism, forestry and other ventures. Then the UK joined the then Common Market in 1973. Its CAP kicked in. Farmers could rest easy. No matter the paper losses, continued subsidies would keep them afloat. Why worry?

Today, in his farmhouse just inside an enlarged Dales National Park – with the Lake District only a few miles westwards – John Dunning, with a lifetime of experience drawn from agriculture and from serving on a range of its representative organizations, has one overriding fear of a "gathering catastrophe": namely, with the disappearance of hill farms, large undertakings such as forestry companies, hedge funds and other well-funded enterprises in the City – eyeing the potential (and profits) from future environmental subsidies and from 'carbon trading' – will effectively create "upland prairies", with few, if any, sheep. The men and women who farmed for generations will have gone and the hills, and uplands of England, will thus have changed forever (Scotland, and Wales, operate under different farming regimes).

Doubtless, some environmentalists will be pleased. After all, the writer and campaigner George Monbiot – scourge of hill farmers – has long complained of our uplands being 'sheep-wrecked'; a landscape effectively trodden down and degraded by tens of thousands of pounding sheep. Much better, he contends, to rewild the uplands and create carbon 'sinks' by restoring peatlands, planting trees and, hence, allowing the ground to absorb water rather than letting it run downhill from hardened grazing areas to the valleys and settlements below, which tend to flood regularly after heavy rain. It is a persuasive argument.

As recorded earlier, the UK has significantly fewer woodlands and forests than other European countries. Nevertheless, buoyed by tax breaks and the prospect of potentially lucrative environmental payments – more accurately, government subsidies to encourage biodiversity – forestation is proceeding apace in some upland areas already, with the consequent removal of sheep. This could be either a threat, or an opportunity for people living and working, in the uplands.

From his 350-acre hill farm in Carmarthenshire, running from 400 up to 1,300 ft in the Cambrian hills, John Lloyd has seen surrounding farms disappear one by one. In their place, tree plantations – "a monoculture of exotic species" – seem to be sprouting all around: "There are several large conifer forests in the area of 20–30 square miles in size," he sighs, as a way of life, stretching back generations, disappears.[6] The Lloyd farm at Cynghordy, near Llandovery, in the family for over a century, has one of the oldest-established Welsh sheep flocks; it's now down from 1,000 breeding ewes to only 200 in the space of a few years. And while John Lloyd insists that while sheep are at the heart of his business, he realized four decades ago that diversification was vital to survive; in 2012, with wife Lucy, he converted old barns into five holiday cottages to supplement a meagre income: "We needed another source … to make this farm sustainable." Interestingly, he has 20 acres of commercial woodland – planted, he insists, to complement his flock rather than to replace it: "Our woodland is run alongside the sheep, utilizing the many benefits trees and sheep bring one another." But there's another important reason, says John: if farmers were more involved with tree planting, forestry companies wouldn't be so dominant in the area – and, of course, the payback would help sustain local economies and diversifying hill farms. As it is, his village of Cynghordy, with 200 souls, is sadly declining as people from England – "in their 50s, with money to spare" – snap up second homes, leaving few children in the village. "That's going to happen increasingly with Brexit as those aspiring to a place in, say, Provence, or Tuscany, turn to places like this closer to home," he predicts.

But there's another side to John Lloyd. He's an ecologist, with a science degree (from the University of Edinburgh), a countryside activist, volunteer with the British Trust for Ornithology – regularly engaged in bird counts – a member of local environmental and farming groups and a lobbyist for 'sustainable' farming. He has seen species – snipe, for instance – return to his hills.

So how does John Lloyd respond to arguments from campaigners, such as George Monbiot, that sheep have degraded the hills to such an extent that the wildlife has been driven out. Sure, Lloyd acknowledges, the hills were over-stocked in the 1980s – but, since then, numbers have steadily reduced to a level that allows grazing to work in harmony with nature and trees. For John Lloyd, in short, arguments surrounding land use – rewilding versus nature-friendly farming – have become "too binary" for his liking.

As it happened, George Monbiot, ironically a fellow ecologist, once lived in mid-Wales – where, it transpires, he coined that now oft-quoted phrase of sheep seemingly destroying the uplands. When, and how, I ask did he become alarmed by the level of livestock on the hills and, thus, become an enthusiast for rewilding? It was when he was writing a book called *Feral* – subtitle: *Rewilding the Land, Sea, and Human Life* – from his then home at Machynlleth, Montgomeryshire. "I think I came up with the term probably in 2011, as a short and simple way of describing what I was seeing all around me and it seems to have caught on a bit," he helpfully responds.

So did one particular event influence him – a walk in the hills, perhaps? He explains that mid-Wales attracted him – "the Cambrian mountains on one side of me and Snowdonia on the other" – because it seemed a wilder place, in contrast to the city. He was soon disillusioned: "The more I walked, the more I found I was seeing less wildlife than in the city from which I'd come." Puzzlement and disappointment turned to despair, he recalls – "almost no trees above 200 metres, you could walk all day and see a couple of crows … no other birds". When he went down on hands and knees in the summer he found few insects, and hardly any flowering plants. Sheep farming, he asserts, was to blame. He looked around the UK, and concluded: "Oh my God, this is universal."[7]

Clearly that's not necessarily an experience shared either by John Lloyd, in his mid-Wales idyll, by the Lake District hill farmer and writer James Rebanks[8] or by other farmers we'll hear from – and certainly not by Phil Stocker, chief executive of the 6,500-member National Sheep Association. Sure, Phil agrees, Britain's sheep industry – by far the largest in Europe – has had a "bad press" recently, particularly in the uplands, home to just under half the estimated 16 million breeding ewes in Britain (a population that doubles with lambing in spring). Phil, who has a small holding in Gloucestershire, blames an "elite group of environmentalists" for the hostility – people, he says, who'd be quite happy to have more industrial-scale mega-farms, and certainly out of touch with reality.[9]

Phil is used to the onslaught directed at upland and hill farmers – the people accused of "sheep-wrecking" and, thus, degrading our highlands – and wonders whether there's a hint of double standards at play here. Criticism of the "elite" exposes a deep ideological divide between enthusiasts for rewilding on the one hand – seemingly happy to embrace mega-farming – and those campaigning for either regenerative or organic farming (including pasture-fed livestock), on the other. It's a debate, in short, between 'land sparers' advocating further intensification of the best farmland, heavily concentrated in eastern England, and 'land-sharers', working in harmony with nature by improving soil quality, stimulating wildlife and wild flowers, often (but not entirely) on a smaller scale through local supply chains. The case of the 'land sparers' is simple: concentrate crop production in tightly defined areas so that more space can be left for rewilding elsewhere. Significantly, figures from Defra, show that large farms – 8% of the total, and largely concentrated in eastern England – grow just over half the domestic food produced annually.[10] Beeswax Dyson and the other large-scale enterprises outlined earlier thus seem to have a seal of approval from some of the rewilders! But the hill farmers – often tenants, like Eric and Sue Taylforth – certainly don't.

The Taylforths have farmed 250 acres among the highest peaks in England for over 40 years. Their 2,000 breeding Herdwick ewes – a small breed of sheep, unique to the Lakeland – roam freely around the common grazings of the surrounding Langdale Pikes, which rise to almost 3,000 ft.[11] Their farm, Millbeck – one of 86 in this National Park owned by the National Trust – has a history dating back to 1621. More importantly, it was one of 15 farms, spread over 4,000 acres, that were left to the National Trust by the writer, illustrator and conservationist Beatrix Potter after her death in 1943. As it happened, she was also a friend of Canon Hardwicke Rawnsley, the first secretary of the Trust and one of its three founders.

This is important to Eric and Sue Taylforth. Why? Because, says Eric, Beatrix Potter had an abiding passion for Herdwicks and for sustaining the farms necessary to continue her legacy – and now, he thinks, those farms face their biggest threat since Potter effectively saved them, beginning with her first farm purchase in 1905. "She was very concerned that mill owners, and industrialists, from Lancashire would buy the farms, and houses, and use them for holiday homes," he recalls. "Now we face a greater threat."[12]

Eric fears for the future through a combination of farm subsidies disappearing and, more importantly, he thinks, a determination of the National Trust to "de-stock" the Lakeland fells, taking advantage of the likely new environmental subsidies on offer from the government – leaving only a rump of Herdwick sheep on each farm – as it pursues another agenda embracing more biodiversity and, hence, benefits from the new ELMS. He points to a farm next door where sheep numbers have been drastically cut following the expiry of a tenancy. "This farm [Millbeck] was built for a reason 400 years ago – the dry-stone walls, and the landscape that went with it – and when all that's gone it will not return." Like other Lakeland farmers, he's out in all weathers repairing dry-stone walls, gates and, increasingly, he thinks, much else besides, "because the [Lake District] National Park has no money and we'll have to do it".

And money – how to survive with diminishing returns and falling incomes – lies at the heart of most problems in the Lake District, in other National Parks and in additional areas of outstanding natural beauty (AONBs), as outlined earlier by John Lloyd in mid-Wales. There, it seems, only those with higher incomes, second homes – which dominate some villages – or retirees new to the area appear secure, while others make do and mend as best they can.

Neighbouring tenant farmers to the Taylforths are similarly perplexed as they view an uncertain future. One, in his early 40s, wonders how he and his family will survive on their Lakeland hill farm without a £20,000 annual CAP subsidy. In England alone, these payments – averaging £27,300 in 2018–19 – account for all the net profits of the average livestock-grazing farm, assuming a surplus is made; often, it is not.

Further north, in the village of Threlkeld on the slopes of Blencathra – one of the Lakeland's higher peaks at 2,848 ft – David Benson is considering his hill-farming future. He owns 60 acres and rents a further 30, with a diminishing flock of 250 Swaledale sheep. Numbers have fallen so he can qualify for a higher-level stewardship scheme – essentially, a government grant paid to farmers who agree to enhance the landscape and encourage wildlife. The quid pro quo lies in reducing sheep numbers, and a consequent subtext: namely, that higher levels of stocking are damaging to the land. David acknowledges that, as one of the smaller farms, he cannot now make a living – "it's not big enough" – and is considering returning to the joinery trade, where he once made wooden gates and fences.

But there's another problem facing hill farmers: a collapse in wool prices, which has further undermined their already precarious incomes. Like others, David Benson remembers a time, not that long ago, when the price he got from selling fleeces – often going for carpet backing and underlay – paid the annual rent for his 30 acres. It's a complaint voiced by many farmers in an industry where wool – rather than meat – was once the primary product produced by sheep.

While the reasons for the price collapse are several and varied – critics accuse farmers of being their own worst enemies – it's easy to forget that swathes of Britain once depended on a wool-based economy. And not that long ago, in historical terms.

Extensive sheep husbandry, linked to the widespread production of woollen cloth, can be traced back to mediaeval times and the monastic orders that accumulated large estates in the north of England for big breeding enterprises. It was only after the Black Death in the 14th–15th centuries, with the loss of rural population, that extensive sheep rearing was introduced in upland areas, where less labour was needed. In the Scottish Highlands, sheep were introduced by estate owners on a significant scale much later, in the 18–19th centuries – at the expense of people being ruthlessly 'cleared' from the land and pushed to coastal areas (or forced to emigrate to North America). By the time of the Industrial Revolution further south in Britain, the woollen industry was helping to power the economy, from Bradford and surrounding West Yorkshire – still a modest centre of wool processing – to the textile towns of the Scottish borders. Mass production required machinery and, hence, factories – and labour, of course, right up to the 1950s and 1960s. Those arguing that the arrival of sheep on the English landscape represents a relatively recent historical phenomenon – with the implication that they are somehow alien creatures – are, thus, wide of the mark.

Although introduction of sheep into Britain is often ascribed to the Romans, the animals are said to have been widespread in the Bronze Age (3000–1200 BC), and by the time of the Iron Age (500–330 BC) wool was spun, woven and became essential clothing. That it isn't today, with the decimation of domestic production, at first glance appears an indictment both of farming collectively and of a wider knitwear industry to address a largely unreported scandal: the dumping, burning and composting of British wool. But, in truth, that's an over-simplified analysis. Who, then, is to blame?

Michael Churchouse, seasoned Devon-based blade shearer and shepherd with a deep knowledge of the industry in the Falkland Islands and in Australasia, has a point when he fumes: "Farmers could do more if they treated wool as something of worth rather than like waste and still expect to get paid for it." He usually spends a month each year in the South Atlantic, shearing some of the 500,000-strong flock in the Falklands, and says the islands' farmers – unlike their British counterparts – know how to respect wool.[13]

Churchouse, who prefers manual clipping to modern electrical shears – and, naturally, has customers nearer to home – is the ultimate evangelist for wool, singing the praises of its many qualities: "It's very adaptable, whether it's warm or cold outside … it's a mini-cooler, with great insulation if it's cold, and it's bio-degradable." The pity, he insists, is that some farmers could do a lot more to keep fleeces relatively clean instead of despatching them soiled – in stark contrast to those he helps overseas.

5.1: Uplands pioneer John Dunning beside his hill farm in the Westmorland Dales

5.2: Grazing in the North Pennines – once earmarked for special status with a rural development board to consolidate farming and diversity the economy

5.3: Grey skies in the Lake District – but for how much longer will the traditional Herdwick sheep graze its fells?

Well, says Jen Hunter, specialist west of England sheep breeder and passionate wool producer, Michael has a point – but, she insists, there's a wider issue to address alongside the inescapable fact that British wool has faced a twin onslaught: the popularity and growth of inexpensive synthetic fibres (and cheap knitwear) from the Far East and the demands of supermarkets for young, rather than mature lamb – which we once called mutton. "We designed our sheep to be super fat and super cheap with the result that wool was side-lined," laments Jen, who has a degree in agricultural and animal science. Lambs, in short, have little wool – and, in any case, thanks to genetics they're not bred to produce fleeces anyway. What she sees as an alarming regression in the industry, away from wool, has been compounded by the apparent disinterest of successive governments: "There's no money for research and development and no national drive to improve the fortunes of wool."[14]

Jen runs 160 acres with partner Andy Wear at Fernhill Farm, in the Mendip hills of North Somerset. The couple have 1,500 Shetland-cross ewes, bred for what they call "fine, colourful fibre and mature meat". Wool is sent to processors in Bristol, Wales, Bradford and elsewhere for washing, spinning, blending, dying and weaving. Some is sold, either to local manufacturers or to their farm, where they make "working-style garments suited to outdoor busy lives".

The couple could tell a sorry tale of a once-dominant domestic wool industry, apparently on its knees – yet they see great potential for expansion and a revival as a real possibility, given more interest from the government. They see no reason, for instance, why other sheep farmers can't deliver wool in good condition to British processers and spinners and, thence, to mills producing cloth for the knitwear and clothing industry. Jen sees parallels with the growing local food movement throughout Britain, sustained by consumers keen to know how and where their food is produced. It's called 'traceability' – and, she insists, a similar philosophy could, and should be applied to wool production. "Just as we ask questions about where our own food comes from, we should also question the source of our wool."

Currently, however, consumers ask few questions about how wool is processed – invariably, cleaning and dying is highly polluting in countries where the UK has 'off-shored' production – and why workers in Far East factories labour on poverty pay in dangerous conditions.

At the margins, a wool revival, of course, would have several positive consequences, social and economic. As well as underpinning and expanding manufacturing, it would – as the Prince of Wales underlined in 2020 – possibly help to sustain the uplands of Britain, home to 44% of breeding ewes, mainstay of the national sheep flock, which can reach 32 million in spring. And, of course, expansion might provide badly needed jobs in rural Britain, where upland farmers are fighting for survival.[15] Speaking to mark the tenth anniversary of the Campaign for Wool, which he founded in 2010, the Prince of Wales underlined the importance to the rural economy of finding better ways to market "this extraordinarily practical and versatile fibre and, at the same time, help so many hard-pressed sheep farmers".

Some of them aren't a happy bunch. Pick up the farming press and a tale of woe emerges from the sheep pen. A farmer from Tunbridge Wells, in Kent, told the *Farmers' Guardian* in late 2020: "We took the first load of fleeces I have shorn to a neighbouring arable farm to be composted. It's now worth less than the diesel I would need to take it to a [nearby] depot in Ashford."[16] Across the UK, it was reported, that farmers were opting to burn or compost fleeces, with prices almost halved due to the COVID-19 crisis. But even before the pandemic, they were low – driven down by a 'shrinking number of [manufacturing] players' and, hence, less competition forcing a further price plunge, according to British Wool, a national marketing organization.[17]

Alarmed by this apparent crisis, the chair of the House of Commons Environment, Food and Rural Affairs committee, Neil Parish MP, took matters into his own hands. He visited a carpet manufacturer – Axminster Carpets – in July 2020, asking them to use more British wool. The company suggested it would certainly help if the product met its standards – but, in truth, the decline of the carpet industry mirrors the running down of wool production generally in the UK.

To answer a simple question – why is so little wool processed from so many sheep on the hills and the lowlands of Britain? – I approached Bradford-based British Wool, a cooperative largely funded by farmers. Its role was explained to me straightforwardly: namely, "to maximize the value of farmers' wool and seek new markets". Easier said than done, it seems. Why? The sector has been hit by several hammer blows.

At a time when multinationals were perfecting synthetic fibres to mimic wool, the UK government further undermined the domestic knitwear industry: in 1992 it scrapped a guaranteed price for all British wool, which helped to sustain production. The impact of this move might be debatable, but it certainly pushed production on a downward trajectory from which it's never recovered. Today Britain accounts for only 2% of wool produced globally, and lies seventh in the world production league; all this in three countries (England, Scotland, Wales) which account for Europe's largest sheep population and one of the highest globally. In the same way that we've surrendered horticulture and soft fruit production to overseas growers, this amounts to another national disgrace, approaching a market failure – highlighted by the obscenity of fleeces being composted and burned because they're apparently 'worthless'.

Is there, perhaps, a circularity in this debate surrounding the use and – let's be frank – the abuse of British wool, and the role it should have in sustaining jobs in the hill farms and remaining mills of Britain – with one part of the industry blaming the other, little attempt at coordination, and still less of an effort to improve quality at the primary source: the sheep farm? British Wool is doing its best, and Prince Charles's campaign has worthy intentions. But surely, in a government once apparently obsessed with a post-Brexit industrial policy, the revival of an age-old industry should be a suitable case for investment and pump-priming – beginning with properly funded marketing allied to adequate research and development to exploit the many and varied uses of a primary product called wool which, shamefully, has been cast as a by-product to a meat now labelled 'lamb' – when, in my formative years, there was only mutton! To be fair, some farmers did spring into action in the summer of 2020; they presented the UK government with a 16,000-signature petition calling for the Chancellor of the Exchequer, and other ministers, to support the case for wool as the ultimate, fire-retardant and efficient form of insulation in a Green Homes grant scheme. Some are even floating the serious possibility of creating small, hard, 'wool blocks' to assist in this process.

West Yorkshire is still the centre of what remains of the UK wool industry – 'although a fraction of what is used to be', according to British Wool. But it has only two scouring plants (which process greasy wool) and one spinning mill supplying the carpet industry, although, around the country, smaller spinning plants, devoted to fabrics, are fighting back, while the famous Harris Tweed cloth, processed and woven on the Outer Hebridean island (which Lewis shares with Harris), has staged a small resurgence. Almost 150 self-employed hand weavers, operating in small sheds beside their crofts (smallholdings), turn out a remarkable range of cloth, all from British wool. It is processed at a main island mill and two smaller ones, where specialist staff dye, blend, card (comb), warp and spin wool for a product protected by its famous orb trademark and underpinned by the Harris Tweed Act 1993. It ensures that the artisan manufacturing process is legally protected and cannot be undertaken anywhere outside the Outer Hebrides, where the industry employs over 300 in total. This gives the tweed, with a global market for its up-market clothing and house furnishings, a status and security other textile manufacturers can only dream about.[18]

But, 752 miles south of the main island town of Stornoway, at the other end of Britain, in the Cornish town of Launceston, the 12 staff at the Natural Fibre Company are certainly not downbeat. They produce yarn from British sheep, through the various stages of production to the final spinning, in a multitude of colours. This is labelled 'vertical integration' in the trade. And in this mill, said to be the most integrated in England, the staff talk boldly of an unfulfilled demand. For Beki Gilbert, who handles sales, it's a "crying shame" that so much wool from the hills and the lowlands is effectively destroyed when there's a growing market for their yarn. "The demand is out there," she contends. The company's recent roots stretch back to 2005, when it received help from the EU to relocate to a 10,000-sq ft factory in Cornwall.[19]

Scroll through the company's objectives on its website, and its support for its primary producers shines through, although – like David Churchouse, the Devon shepherd and shearer – they think those who raise sheep could keep the fleeces in a better condition. But the small company is adamant, in its aims and objectives, that maintaining the pastures of Britain is key to its future: 'We believe there is an important future and role for farming and for the countryside as the lungs and heart of the UK.' As such, it's committed to supporting fleece suppliers with advice on 'getting their best fibre from their flocks'.

That, of course, raises further questions about – let's be blunt – the comfort zone under which the EU's CAP has feather-bedded farmers, from the wealthiest with the largest holdings – who get the most – to the poorest on the hills and uplands, who depend on it totally because it represents well over half their income. Some, like John Dunning from his home in the Westmorland Dales, have long been warning that those farming the hills have been living on borrowed time without an alternative income. Robin Milton, who farms just north of the Natural Fibre Company on the Devon–Somerset border, knows the arguments all too well.[20]

Robin's family have farmed on Exmoor for eight generations. With his brother and family, he has a mixed livestock farm – 800 sheep, 300 cattle and a herd of the famous Exmoor ponies – on 1,000 acres, with further access of 3,500 acres of common grazing on surrounding moorland. Although EU subsidies provide the family with much of their income, Robin is adamant that the system is not only wasteful but has also outlived its usefulness. He thinks EU subsidies have disempowered farmers with inflexible rules under a tick-box culture in which "we've lost ownership of what was there". Hills, he acknowledges, have been over-stocked, largely at the behest of the EU, which, at one time, offered a 'sheep premium', which served only to keep numbers high. But since 2005 he's seen numbers drop by 40%. Farmers in his area are now working with Natural England to improve biodiversity, restore peatland and – successfully – encourage wildlife to return.

As such, leaving the EU has been turned from being a threat into an opportunity for the 270-square mile mosaic of pasture, rolling moorland and rugged coastline of the Exmoor National Park, which straddles Somerset and Devon. A range of groups, from a well-established hill-farming network, the National Park authority – chaired, conveniently, by Robin Milton – Natural England, the Royal Society for the Protection of Birds (RSPB) and local landowners, combined four years ago to forge a post-Brexit future. Labelled 'Exmoor's Ambition', its aim is to chart a practical future for the National Park, farms and businesses tied to the land began long before the government began seriously considering the issue, around a central theme: how to value nature and biodiversity and put a price on natural assets? Robin Milton happily requotes Emma Howard-Boyd, chair of the Environment Agency that "you will get the environment you pay for". The now formalized Exmoor's Ambition group has been arguing that farmers should be offered financial incentives to produce 'public goods' such as improving landscapes, encouraging flora and fauna, preserving historic buildings and creating carbon sinks in renewed peatland. To assess progress, and justify a public subsidy, they stress that farmers should have to sign up to an environment agreement and face regular monitoring. But could they survive the withdrawal of an annual subsidy, with a still-to-be-determined replacement which will clearly never match the level of support from the outgoing CAP regime? A process of attrition, then, I ask tentatively, with casualties along the way? "I fear that is possible, and farmers are a pretty tough lot ..." replies Robin Milton.

As John Dunning, life-long uplands champion and entrepreneur, recalls, the future could have been so much brighter today if opportunities to diversify were grasped 50 or so years ago – when it was clear that the prosperity brought to lowland farming by the post-war subsidy regime could never deliver the equivalent benefits to the uplands. Loss-making hill farming, in short, has survived in scale only through generous subsidies – invariably, at some cost to the environment.

Whether many hill farmers become de facto land managers, dependent on the uncertain revenue flowing from a nature-based economy, is anyone's guess. Will it prove a giant leap into the unknown, a step too far for many? Will others seize the initiative and embrace the only alternative, which might soon be the new normal: working in harmony with nature and getting paid, however modestly, to nurture the landscape? Some have already taken up the challenge with enthusiasm.

6

The Climate Challenge:
Land versus Water

And the horizon stooping smiles / O'er treeless fens of many miles.[1]

To see the tidal defences and inland channels regulating water and protecting farms and communities is to marvel at the monumental task of transforming 1,500 square miles of wetland into the country's best arable acres. With a habitat and history quite distinct from the rest of the country, we call this huge area the Fens: drained from the early 17th to the mid-19th centuries, at some cost to wildlife, nature and to a distinctive way of life for its people.

For Daniel Defoe, it was 'all covered in water like a sea ... the soak of no less than 13 counties':[2] a giant sponge absorbing the water flowing into a landscape, once one of Europe's great deltas and 'most diverse environments'.[3] Today it is interlaced by a network of straightened rivers, parallel channels, small reservoirs, endless long and high embankments, several hundred pumping stations and sluices to hold back water; all this to protect vulnerable and drained farmland, semi-rural communities and, further upstream, Cambridge itself.

Now these flatlands, a food bowl for Britain, often below sea level and leaching carbon into the atmosphere, are degrading as the ground sinks – and the country faces tough choices. Should part of them be returned to a natural state of meadow, marsh, meres (lakes) and meandering rivers – in short, 'rewetting', to contain carbon – with the remainder reinforced to protect farmland at a cost of billions over the 21st century? Will the government, when it finally considers the issue, authorize limited funding and make do and mend as best it can? Or are the Fens destined to eventually become a wetland once again, by default if not by design? As we shall see, tentative signs of action are emerging, with a new government task force charged with addressing the challenges in our lowland peatlands – none bigger than the Fens.[4]

One historian who has undertaken a recent detailed study of the area is in no doubt that, even under conservative climate forecasting, sea level rises – perhaps by a third of a metre by 2050 and between 1 and 2.4 metres within 100 years[5] – are likely to lead to the 'seasonal or permanent inundation of much of the Fens without the construction of sea walls of prohibitive size and expense'.[6]

As Britain gets warmer and wetter, putting scores of coastal communities in the Fenland and further afield at risk, the Environment Agency (EA) in England appears in no doubt that the country has approached a tipping point: invest heavily in climate adaptation and mitigation or accept the consequences of a rapidly receding coastline, and the need to relocate vulnerable communities inland.[7] Even so, in a country with 'some of the fastest eroding coastline in Europe', up to 115 miles of it would still not benefit from a wider shoreline protection programme estimated by the CCC to cost £18–£30 billion this century.[8]

Combine that with the added threat to valuable farmland, through degradation of peatland, with all the implications for domestic grain and other crops in the Fens and beyond, and you see the problem: on one hand, as the NFU helpfully underlines, the Fens include half of the nation's best (Grade 1) arable acres and produce a third of its fresh vegetables alone; on the other hand, many of the flood- and water-management structures in the Fens are 'coming to the end of their design life', and therefore 'significant investment is needed', according to the EA.[9]

This is compounded in an island nation vulnerable to rising sea levels as climate change undermines an already low, and falling, self-sufficiency in food and the vulnerability of the east coast to rising sea levels; by contrast, the Netherlands, arguably more at risk, with considerable land below sea level, has world-beating sea defences.

Several issues – aligning food production with nature's renewal to address the climate emergency, for a start – need tackling in the round by active government. Technology can help by making better use of cultivatable land away from the Fens, sometimes occupied by livestock, expanding large-scale, indoor crop production using water rather than soil – known as hydroponic growing, or vertical farming – and, possibly, switching other crop production to 'wet farming', or paludiculture, in places like the Fens. However, can all these alternatives – some contested as impractical and untested – compensate for the loss of Fenland farming?

This is where the new task force on 'sustainable farming' of lowland peatlands – doubtless, a contradiction to some – comes in.[10] It has the challenge of squaring a circle: namely, how farming in the Fens can be balanced with reducing carbon emissions. All that is against the background of this reality from Defra: 'Centuries of draining these areas to support intensive agriculture have led to degraded peat soils which emit nine million tonnes of greenhouse gases each year.'[11]

The CCC, a government advisory body, has put this into stark relief: although lowland areas, notably the Fens, account for just 14% of UK peatland – with the uplands containing the rest – they account for 56% of all peatland carbon emissions.[12] To say that the task force has its work cut out is an understatement. But at least its report might drive home some unpalatable truths: namely, Fenland and farming working in harmony involves tough choices.

All this, of course, is against the background of the climate emergency: wetter winters and much hotter summers. We have already been warned. Floods, in the 2010-20 decade, should have been a wake-up call. First, an alarming North Sea storm surge in late 2013, inundating coastal land. That was followed by ultra-heavy downpours in the Pennine towns, and elsewhere, on Christmas and Boxing Day in 2015. By February 2020, storms Ciara and Dennis, in quick succession, wrought flooding havoc on parts of Herefordshire, Shropshire, Worcestershire and the South Wales valleys. Then, in late 2020, the Great Ouse, feeding into the Fens, burst its banks in Bedfordshire, causing extensive flooding – and leading to near-panic in Fenland and anxious moments for those manning flood defences. Briefly, as ever, the country took notice – and, to be fair, the government in its 2020 budget did promise to double spending on flood and coastal risk management in England to £5.2 billion between 2021 and 2027.[13]

And while all this was happening, the steep-sided Calder valley, in West Yorkshire, was presenting further challenges, distinct but no less threatening than the continuing problems in the Fens: how to safeguard thousands of people, and countless businesses, in the towns of Mytholmroyd, Hebden Bridge and Todmorden: the extreme example of flood-prone Britain, exacerbated by the misuse of moorland above the valleys, and replicated to varying degrees in other places.

With extensive flooding in 2012, 2015 and then in early 2020, these Pennine valley towns are in the highest risk category and, in the worst case, are likely to be overwhelmed by water tumbling down steep hillsides. All this, despite monumental efforts by the EA, Calderdale Council, the National Trust and an impressive local charity, Slow the Flow – with a small army of volunteers and flood wardens on hand to sound the alarm if necessary – to tackle flooding at source below the grouse moors, and the valley now being safeguarded with new and emerging multi-million-pound defences.[14]

While those moors were drained under a government scheme from the 1950s to the 1980s to improve grazing for sheep, it also made them more attractive for what the field sports lobby casually call 'game birds' – although you might think that this 'sporting' label appears the antithesis of harmless exercise. The preparation for shooting – burning moorland to encourage the growth of new heather favoured by grouse – has its consequences: blanket bog, with its rich mix of small shrubs and sphagnum, has reverted to what some call a monoculture of heather and drier upland peat. This leads to intensified water run-off from the hardened and less-absorbent ground, down to the valley floor – compounding the flooding risk in the Calder valley towns.[15]

Left alone, peatlands continuously accumulate carbon in their natural, soggy state, while storing water and – crucially for valley settlements, like the Calder valley – slowing the flow and helping to 'alleviate the risk of downstream flooding', according to the CCC.[16] Grouse shooting, thus, has profound implications beyond the drained and dry moors.

The fight for Fenland

Draining the Fens, a considerably more ambitious exercise spread over centuries, has consequences too with, in the CCC's words, 'historic and ongoing drainage resulting in significant peat loss and shrinkage'. There's no reason, then, to assume that Holme Fen, in Cambridgeshire – just one slice of Fenland – is exceptional: 100 years of drainage has resulted in an estimated four metres of peat shrinkage.[17] The statistics speak for themselves; maybe protesters, opposing drainage centuries ago in what amounted to a near-insurrection, knew something which the aristocracy chose to ignore.[18]

While Fenland drainage began over 400 years ago, with improvements and extensions over the centuries, work was often delayed by violent protests against ruthless land-grabs – ostensibly legal – which forced locals from common land. Across England, a quarter of all cultivated acreage was appropriated by politically dominated landowners dispossessing the long-held grazing and cultivation rights of people working the land: 'an extraordinary effort, unique in Europe, to enclose common lands by decree'. And none more so than in the Fens, where the 'destruction of the common marsh would take away the economic independence on which the people's freedom relied'.[19]

By fighting back, the Fenlanders, at the very least, held the line for a time against powerful interests bent on dispossessing them. By a twist of fate today, parts of the very area transformed from wetland to farmland are being highlighted as a suitable case for 'rewetting' and, hence, returning to the state enjoyed by earlier, dispossessed commoners – in short, no longer resisting the forces of nature, as peatland is restored and 'carbon sinks' are created to address the climate emergency.

Spread largely around a broad, rectangular bay on the corner of East Anglia – the Wash – the Fenland was once a rich, wetland habitat where small-scale farming happily coexisted with the natural flow of nature: incoming tides meeting outgoing fresh water, stimulating rich pastures, endless reed-beds and plentiful wildfowl, fish and eels. In eastern England, the task of draining a mosaic of meandering rivers, creeks, small islands and salt marshes – rich in vegetation and wildlife – began in earnest under the leadership of Cornelius Vermuyden, a civil engineer from the Netherlands-cum-honorary Englishman.

Subsequent improvements over the centuries have created the distinctive flatlands of today: a corner of the country largely out of sight and mind to many who, unknowingly, depend on this region for much of their food. With 57% of our crops coming from just 8% of the country's largest farms, we can safely say that many of those farms are either in the Fens or on its periphery.[20]

By the end of this century, the EA is quite clear that rising sea levels, caused by changing weather patterns, could lead to 'a permanent loss of land', with some of the most 'productive and higher value' acres on flood plains needing to be 'freely drained and defended from inundation'. The EA says 'difficult issues' consequently need addressing in low-lying catchments like the Fens, where 'premium food production' is dependent on flood-risk management and land drainage.[21]

Critical Fenland?

Today, this fascinating, flat landscape of endless cultivation, quite distinct from the rest of the country, and modernized over the centuries, is central to Britain's self-sufficiency in food. It represents critical infrastructure for our three nations. In the catchment of the Great Ouse alone all this is sustained by an 'extensive and complex network of assets' including 138 pumping stations, 24 sets of sluice gates, 60 miles of coastal defences and 252 miles of inland embankments protecting farmland, often below sea level. None is more crucial than a seemingly modest structure across the Great Ouse near the market town of Downham Market.[22]

For the EA, charged with water management and flood prevention, Denver Sluice ranks in strategic importance close to the Thames flood barrier – but, rather than just safeguarding people and property, the large gates set in stone arches protect 457,000 acres of farmland, 131,000 households, 13,200 industrial and commercial properties, many villages, market towns and, at the extreme, even the university city of Cambridge upstream and its valuable bio-tech sector. The sluice has failed twice: first in 1713, when its original gates were lost in an exceptionally high tide, and again in a 1953, when a storm surge burst through the gates, leading to floods inundating 400 square miles of the Fens and nearby land.[23]

First built in 1652–53 by Vermuyden and his team, the sluice has had various incarnations. It collapsed and was rebuilt on several occasions, and was subsequently improved in 1982–86 when the flood gates were replaced; IT gadgetry was later installed. But, remarkably, it still depends entirely on human endeavour and the watchful eyes of its superintendent, Dan Pollard, and a colleague, who both live on the site, 12 miles inland from the port of Kings Lynn.

Operation of this small complex appears deceptively simple: at high tide the three large Denver gates are closed, keeping the sea out and forcing water back up the river channels; as the tide recedes, the gates reopen. The sluice thus manages the Great Ouse, its tributaries – the Cam, Lark, Wissey and Little Ouse – and, in total, 620 miles of rivers upstream. It's a sophisticated operation. If it rains heavily, with a likelihood of flooding, Dan can regulate the flow, and the depth upstream by several metres. A network of ultrasonic boxes at key parts of the river system – such as Ely – calculate water depth through soundwaves and send the findings to the control room at Denver. For two days over Christmas, 2020 – 24–25 December – river levels rose to the highest in 22 years at Ely – 15 centimetres from breaching defences – as floods hit the Great Ouse further upstream in Bedfordshire. Dan was working 12-hour days to keep the Fens safe. If necessary he can utilize nearby overflow courses, not least a 12-mile channel parallel to the Ouse built after catastrophic floods in 1947 – effectively a long lake which can almost double in volume to take excess water, rising an additional 2.5 metres above its average 3-metre depth . If the weather turns for the worse, he's up all night.

Dan, who has been at Denver for 15 years, keeps a constant eye on a screen in his operations room, to where those vital ultrasonic monitors upstream feed information on river height and conditions in the giant catchment area of the Great Ouse. If conditions demand, it's round-the-clock working. And in the last decade, he's noticed one significant climatic change: the greater intensity of rainfall, and the localization of heavy showers. This can, for instance, lead to alarm signals of high water levels on the Great Ouse at Ely, 19 miles away, flashing on his screen while Denver remains dry. Within an hour of Dan working his magic by diverting water, the water level at Ely can be reduced by a metre.[24]

But the Denver Sluice has another purpose: defending the Fens from the ravages of the North Sea. This is no distant, theoretical threat. In December 2013, for instance, a storm surge – a combination of spring tides, low pressure and gales from the north east – proved as violent as the great flood in January 1953, which engulfed the east coast. Then, water levels rose to six metres above normal and 300 people lost their lives in Suffolk and Essex. Mercifully, in 2013 no lives were lost, although a vulnerable coastline was further eroded,

That December 2013 surge remains vivid in Dan Pollard's memory; he recalls water lapping over the top of Denver's flood gates. For a time, his team held their breath. But defences, much improved since 1954, held firm … just. What if they hadn't, I ask tentatively. "Catastrophic," replies Dan.

Today those defences in the Fens, while a shadow of the network constructed in the Netherlands after that 1953 North Sea storm surge, are still holding the line – while the EA presses for a longer-term commitment to bring its network up to date and withstand the pressures of high winter rainfall, increasingly hot summers, rising sea levels and farmland which sinks further each year as the peat evaporates and river levels consequently rise. The assessment of experts close to the ground is sobering: 'Much of this infrastructure is nearing the end of its design life and will soon require significant investment.'[25]

The Fens – and the coastline of eastern England – now face some tough choices when, and if, the government begins addressing the climate emergency with detailed plans, rather than generalized statements leading to vague, unsubstantiated commitments about meeting a zero-carbon commitment before 2050. "We've got time, but by five to ten years we need to get on the front foot and step up to the epochal challenges," one senior official in charge of flood resilience and remediation acknowledges. Recalling substantial government investment on the back of "catastrophic" flooding in 1947, costing several billions in today's money, he adds despairingly: "How do you stimulate urgency before disaster?"

Jim Hall, Professor of Climate and Environmental Science at Oxford University – lead author of a CCC 2018 report into coastal communities – is similarly concerned about inaction at the top and says the government "aren't preparing anything you might describe as a plan, although some civil servants get this [agenda]'. But he does point to difficult choices: arguments, for instance, on spending relatively large sums to protect smaller coastal communities pitted against the stronger economic case for defending a vast area of the Fens which contain the country's most productive land.[26]

But here again, within the Fens, tougher choices are presented: continue farming on the drained land, or return it to a wetland state, however partial. On the one hand, says Hall, the Fens' importance as a food bowl for Britain should not be underestimated. But, on the other hand, 'there's more of a case for restoring them because farming inevitably reduces the amount of carbon in the soil ... on the other side of the balance sheet you've got the added benefit of reducing carbon emissions, preserving peat, and these would be really exceptional, low-lying, wetland areas'.

This is no longer a passive debate. In another sobering report, in June 2020, the EA pulled no punches. After raising a range of options, it soon becomes clear that 'managed realignment' could emerge as the most practical and cost-effective plan. That means accepting that some of the coastline will be left to erode, with other parts strengthened while, inland, a 'nature based solution', through salt marsh restoration, could be the way forward for part of the Fenland, thus creating carbon-friendly peatlands once again by allowing the sea to penetrate defences while still safeguarding some farmland.[27]

You get a sense of this future, albeit on a small scale, by the estuary of the River Aln, in Northumberland, near the pleasant coastal village of Alnmouth. Overlooking the county's spectacularly wide beaches, the EA has worked with landowners to remove flood banks, which once protected pastures from the sea, and rewet the estuary. On a large display board it proudly proclaims: 'Now sea water regularly flows over this land and salt marsh plants [glasswort, thrift, for instance] have become established.' Mudflats and salt marsh creeks now support invertebrates: food for a wider range of wading birds and wildfowl, such as lapwings, widgeon, teal and curlews.

Could this presage a restored Fenland? On a much larger scale, the 2,000-acre Wicken Fen, near Ely, is a showcase for the brave new world of rewetting, although it's currently only one of four small fragments of Fenland in a natural state. Run by the National Trust, it's one of Britain's oldest nature reserves – 'extremely precious and rare in a national, and international context' – with a history stretching back to the late 19th century, when the banker Charles Rothschild, a keen entomologist, donated a chunk of Fenland – since expanded by substantial acquisitions. With restored reed beds and marshes, wildlife which disappeared long ago – cranes, bitterns, for instance, along with snipe and grey partridge – is returning.[28]

This, then, represents rewetting, or the relatively new concept of 'rewilding', on a fairly grand scale. It provides an insight into a contested area: the case for restoring a wider block of Fenland to a natural state – and, for the National Trust, Wicken Fen undoubtedly provides a showcase for others to follow.

The Trust has a vision for recreating a 'diverse landscape' from Wicken Fen to Cambridge, 17 miles away. Sarah Smith, its local manager, talks enthusiastically of "eco-system services" – an area which, she acknowledges, is "little talked about" because the wording mystifies many. She explains this more simply as "locking carbon away in the soil" and, thus, helping the country meet its target of net-zero emissions before 2050. That, of course, can only be achieved by returning the Fens to a natural state because, as a colleague helpfully elaborates, 'draining them has produced greenhouse gases at a phenomenal rate".

From the government's advisory body, CCC, to its delivery organization, the EA, it's now clear that what once was considered a distant possibility – rewetting – has crept up the policy agenda, underpinned by the prospect of soils eroding and peat oxidizing in the Fens. In the words of one official, echoing the findings of the CCC at Holme Fen, this can have only one outcome if allowed to continue, namely: "The farmland [in the Fens] is going to disappear anyway ... so far better to plan for returning it to a natural state."

But change, inevitably, prompts opposition. No surprise, then, that when the vision for expanding Wicken Fen first surfaced, an online petition emerged opposing the plans on the grounds that valuable agricultural land would be lost, while visitor traffic to the National Trust's reserve would increase. As it happened, a counter-petition, supporting the Trust's scheme, attracted more signatures.

Yet, viewing the infrastructure created to manage water in the Fens and sustain farming, is as impressive as it is sobering. The sluice at Denver might be the ultimate manifestation of Fenland's (relatively recent) physical resistance to the sea and of the management of water – but, beyond it, there's 3,800 miles of watercourses, plus a network of internal drainage boards. When government advisers mention a confusing, rather than cohesive, organizational picture in Fenland – EA, Natural England, county councils, district councils, drainage boards – you know what they mean.

But, in truth, there are no easy solutions as the debate around the future of the Fens rumbles on: partly surrendering them to the sea, along with some associated coastal defences, or defending them for the foreseeable future. And a glimpse of the task facing the EA came late in 2020 when a senior EA official underlined the Fens' vulnerability at a meeting of councillors from the Kings Lynn and West Norfolk district. This is an area most at risk from flooding and from high tides and storm surges – with the town of Kings Lynn itself particularly vulnerable. The official reminded the gathering of one chilling statistic: by the end of the century, two-thirds of the Great Ouse Fen – sustained by the Denver sluice – will be below sea level.

As the meeting drew to a close, one councillor asked the leading question: namely, moving people away from areas most at risk. "That's adaptation, absolutely," the EA official responded. The discussion then promptly ended. "I think we've got a subject for another day there," the chairman concluded, as though this hot topic was off limits.

For the EA, it certainly is not. The coastline is eroding faster in Britain than elsewhere in Europe, with land disappearing at a rate of four metres annually in some places. By the 2080s, according to a worst-case scenario by the CCC, more than 100,000 properties, plus roads and railway lines, may be at risk from erosion.

Towns and villages on the edge

It's December 2013. High above the dunes on the Norfolk coast, villagers in Hemsby gather in the local pub – ironically, for a fundraiser to support better flood defences for a community literally living on the edge. Soon all hell breaks loose outside the Lacon Arms. Gale-force winds whip the shore, sending 20-ft waves towards the sand cliffs supporting houses. A bungalow, 300 feet from the pub, soon slips into the sea. Six other houses follow, along with the local lifeboat shed. Ian Brennan, chairman of the Save Hemsby Coastline group, formed to protect the ultimate at-risk village, recalls: "Up to 50 of us went out and tried to get people out of houses being washed

6.1: Big skies: geese over the Great Ouse, a principal river of Fenland
Source: Paul Burrows

6.2: Denver Sluice: critical infrastructure in the Fens to protect the country's best farmland, much of it below sea level
Source: Paul Burrows

6.3: Fenland
is maintained
by a complex
network of
sluices, pumping
stations, and
relief channels
to take excess
water
Source: Paul Burrows

6.4: A sign
of the future
for the Fens?
Rewetted salt
marsh in the
River Aln
estuary,
Northumberland

6.5: How it
should be:
well-spaced
woodland to
encourage plant
life, rather than
dense forests of
alien conifers

away ... earlier in the day [they] were 40–50 feet away ... it was like a disaster movie."[29]

Along the coast of eastern England, from the Thames to the Tyne, it was soon clear that the country had experienced a tidal surge to equal the devastating North Sea storm of 1953 which cost 307 lives in England and 1,835 in the Netherlands. While the topography of England is distinctly different to the Netherlands', where a world-beating flood defence system now keeps the country relatively secure, swathes of our coastline, river systems and inland plains remain alarmingly exposed to rising sea levels – which have risen by 15.4cm since 1900 and are expected to rise by at least a metre from current levels by 2100 – and increasingly heavy rainfall inland, as the residents of the Calder valley, in West Yorkshire, can attest.[30]

Hemsby and a string of other villages, might appear to be on the front line – but, arguably, they're not quite as vulnerable as the Gwynedd village of Fairbourne, on the Barmouth estuary in West Wales. Here people have been labelled the most likely to become Britain's first 'climate refugees', following warnings from Gwynedd County Council that they face evacuation by 2054 because coastal defences can no longer protect a population of 1,000. In officialese, Fairbourne will be 'decommissioned'. The future of other settlements is on the line. It's easy to see why.[31]

On Britain's fast-eroding coastline, land is disappearing by four metres a year in some areas, particularly in the east of England. As a result, our nations – England, Scotland, Wales – face two challenges: reinforcing sea defences where practical and, elsewhere where it's impractical, preparing to relocate residents and businesses. Remarkably, there's no plan to deal with these challenges, no data on how many houses have already been lost and no collective estimate – among those local councils monitoring coastal regions and erosion – of how many homes may be vulnerable.

Jim Hall, Professor of Climate and Environmental Science at Oxford University, calls this a "ticking time bomb". In arguing that existing plans to protect coastal areas are unfunded and unrealistic, he estimates about half of England is protected by defences, such as sea walls and promenades, which are reaching the end of their natural lives, and that maintaining them will be unaffordable.[32]

The warnings from the CCC about the need for openness and urgency when dealing with the reality of eroding coastlines and threatened communities –'we have to start those difficult conversations now' – are shared by the Town and Country Planning Association (TCPA), which has long complained about the absence of both a strategy and coordination among a myriad of organizations and drainage boards in addressing rising sea levels, and the consequent threat to coastal communities, along the east coast. It has called for new legislation – a National Resilience Act – to create special development corporations charged with both addressing flood prevention and planning new settlements to replace those lost to the sea.[33]

Already the EA, which operates in England – Scotland and Wales have separate, broadly similar bodies – accepts it will be impossible to safeguard all areas currently protected by sea defences. And it warned in 2019 that 'retreating' from some higher-risk areas – so far not identified officially – could be inevitable in the face of rising sea levels and increasingly extreme weather. But, so far, there's neither government money allocated for moving people and communities to safer ground, nor an overt acknowledgement of the problem.

Warnings ignored?

That official indifference similarly applies to farmland. In 2010, for instance, a report from the Government Office for Science, *Land Use Futures*, noted that while 2.5 million acres – or 9% of our agricultural area – occupies flood plains, it embraces our best land: 57% of Grade 1 agricultural land and 13% of Grade 2 alone, 'an important asset in terms of national food security', according to the report, which put the capital value of 'at risk' land at £15 billion.[34]

The management of flood alleviation, and drainage, is therefore critical to maintaining domestic food production. The choice in 2010 was put starkly in the report: either governments provide high levels of flood protection to resist rising coastal threats, 'realigning defences', or they abandon large tracts of land to the sea. Presciently, the report warned of the potential threats from rising sea levels and storm surges – three years before the North Sea surge of 2013. If that wasn't enough, in 2014 the National Audit Office warned that millions of acres of farmland in England are potentially at risk from flooding because government spending on flood defences has been spread too thinly.[35] Few people listened (and one member of the Foresight team told me they had been told to – would you believe it? – 'water down' their report before the 2010 general election).[36]

We are left, then, with one inescapable conclusion: invariably, only when a storm strikes, the heavens open and large tracts of land (and countless houses) are flooded do we – and governments – sit up and take notice, by which time it's often too late. But not, as we shall see, in the Calder valley, where an alliance of organizations – Calderdale Council, the EA, the National Trust, Yorkshire Water and an impressive array of volunteers and charities – have taken the initiative to improve the resilience of this delightful corner of West Yorkshire's hill country. The 'blame game', directed at governments local and national, which invariably follows a deluge in other areas – 'why don't *they* do something?' – is refreshingly absent.

Sure, the aftermath of two storms in the winter of 2020 – Ciara and Dennis – did prompt calls for action from several quarters, not least from the TCPA. In a letter to the Prime Minister it complained of the absence of any plan to deal with the impact of at least a 1.15-metre rise in sea levels by 2100 – with eastern England clearly in the firing line. The absence of a response was deafening.[37]

Talking to the people who farm the land, and the engineers and hydrologists responsible for keeping it secure – often below sea level – one issue becomes clear from the outset: the fragility of the Fens, plus other land, and, bluntly, how close some places are to being overwhelmed by the forces of nature unleashed by global heating and climate change: wetter winters, hotter summers, and the ever-present danger of another tidal surge. And here's the irony: in an area where the EA and a network of internal drainage boards are constantly fighting a battle to contain Fenland water, farmers (as earlier noted at Elveden, Suffolk, the country's largest arable farm) highlight water shortages.

These concerns have been underlined by Sir James Bevan, chief executive of the EA, who cautions that hotter and drier summers will mean water shortages by 2050 – particularly, he says, in south-east England, "where most of the UK population lives". That's small comfort to Essex, which draws some of its water from the Great Ouse catchment. No surprise, then, that people close to the action in the EA question how long the Essex connection can realistically continue. As Sir James Bevan has cautioned, on current projections many parts of the UK will face water shortages by 2050, as a result of hotter and drier summers.[38] As his agency has noted, this falls into the broad argument for 'significant investment' in flood and water management, 'balancing the needs of people, the environment and agriculture'.[39] Tough choices indeed.

To varying degrees, those charged with safeguarding the infrastructure of the Fens, advising the UK government and, hence, speaking truth to power – the EA in England, and the CCC – are quite clear where priorities should lie: hopefully, a commitment for multi-billion-pound investment this century to safeguard national assets and protect, and relocate, communities. This is no longer an academic debate.

Flooding fears

It helps, of course, when communities themselves decide to take matters into their own hands. That's what happened after extensive flooding in the Calder valley and beyond on Boxing Day in 2015 – so severe that it's still labelled "apocalyptic" by Scott Patient, a local councillor. "The valley has always flooded, but the magnitude has increased in the last 20 years – it's now ever more extreme – but we've never seen before the likes of that 2015 event" (when his own house was flooded), he recalls. It's a measure of how seriously Halifax-based Calderdale Council takes the issue that Patient – passionate in driving forward "natural flood management" in the hills above his village – sits on the council executive (or cabinet) with responsibility for climate change.[40]

Late 2015 proved a defining moment. It prompted a group of locals from three towns (Mytholmroyd, Todmorden, Hebden Bridge) and beyond to launch 'Slow the Flow', which has grown into one of the country's largest charities devoted to safeguarding communities from flooding. It's a partnership blessed with expertise from a range of local professionals, backed by around 1,000 volunteers.

With 2,781 houses and 4,416 businesses flooded as water cascaded from the Pennines into towns on the valley floor on that Boxing Day – some localized rainfall exceeded over 100mm in 24 hours – locals decided to take immediate steps. Bede Mullen, a retired academic and engineer, who chairs the Slow the Flow charity, acknowledges they can never replace what he calls "hard engineering solutions", "but we can reduce dependence on them at negligible cost". He recalls what while the floods had a devastating impact on "lives and livelihoods", it became clear from the outset that "there are things we can do immediately to reduce the impact of flooding" through "natural flood management" – or, literally, slowing the flow of water down the steep hillsides. In five years, the group have created 619 small 'leaky' dams, as well as eight large water-storage areas and laid hundreds of felled trees along the contours of the hills to divert water. At the same time, the partnering EA is providing the 'hard engineering' in one of its highest-risk river catchment areas, heavily populated, where the valley accommodates a railway line, a major road and a canal. Since the floods, for instance, houses have been demolished at Mytholmroyd to provide room for widening the River Calder by around eight metres, a new bridge has been built over the river and defensive walls erected, at a cost Scott Patient puts at £40 million. Nearby towns are in line for other measures; preliminary work on a £30 million project is under way in nearby Hebden Bridge.[41]

Early, in 2020, the heavens opened again above the valley, flooding roads, houses and the local primary school in Mytholmroyd, although this time new flood defences protected previously vulnerable properties. But looking to the moors above, Scott Patient, his council colleagues and the Slow the Flow volunteers doubtless reflected that, while their collective actions on the hillsides have played a part in reducing the flooding risk, the pursuit of grouse – and the destructive burning of peatland to serve a tiny minority of shooters – is impeding their valiant efforts. Bluntly, when peat bogs are restored and vegetation takes root again, water is absorbed and, consequently, held back – thus reducing the flow down the hills to the Calder valley. The target for their ire is the owner of Walshaw Moor, a 6,500 acre 'sporting' estate – and the case for rewilding places like this, and addressing natural flood management in the round, becomes more compelling by the day. Scott Patient, for his part, is sure of one thing: "Their time is coming to an end; the writing is on the wall."[42]

The climate challenge

In examining the pressures and the challenges facing our island, driven largely – but not entirely – by the climate emergency, it's important to stress that life for Britain outside the EU should offer significant opportunities to rethink and redesign our approach to that most basic resource: our land.

Some issues – warmer summers, wetter winters, rising sea levels – may seem beyond the control of humans. Through mitigation – defending some places downstream, and along the coast, from flooding; accepting that others, regrettably, face an uncertain future – we are already charting a realistic way forward. But, through adaptation, we can make a big difference and begin framing a more secure future. Think, for instance, how much safer communities alongside the River Calder would feel if the grouse moors above were returned to a soggy peatland state, to complement the monumental voluntary effort by locals to slow the flow on the hillsides below. A combination of nature-based solutions alongside expensive hard engineering, from flood gates to raised, strengthened banks and widened rivers, should provide what the experts call a 'whole catchment' approach, beginning with addressing flooding at the very source – on the moors. That means progressive land management and reworking the uplands to address the climate emergency. In the words of Sir Ian Boyd, former chief scientific adviser at Defra, "We need a large, radical, transformation and we need to do it quickly in the next decade."[43]

Professor Boyd believes that half the nation's farmland – mainly the uplands and hills – should be transformed into woodlands and natural habitats to fight the climate crisis and restore wildlife. To achieve this, he says the numbers of sheep and cattle would need to fall by 90%, with farmers instead being paid for storing carbon – thus helping prevent floods while providing renewed landscapes for recreation.

The analysis from Boyd, who spent seven years at Defra, rests on a simple premise: that half of farmland – hills, uplands and lower pastures – produces just a fifth of the UK's food and would be better used to deliver 'public goods'; growing more trees, providing better water, creating more diverse habitats and storing more carbon. And farmers, he thinks, are potentially "sitting on a goldmine" if subsidies – through the emerging ELMS – are based on delivering these public goods by valuing nature.

As it stands, only 15% of UK agricultural land is being used to grow crops for human consumption, with a further 22% delivering crops to feed cattle and sheep; grassland used for grazing livestock accounts for the remaining 63% of farming land.[44]

It's true, as Boyd argues, that most livestock production in Britain would be unprofitable anyway without public subsidy through the EU's CAP – and, consequently, many hill farmers and some on the lowlands, are living on borrowed time. But there's a real danger here that many, often tenant farmers, will take a likely 'decommissioning' or exit payment from Defra and leave the land – allowing existing landowners, and maybe other enterprises, to sweep into the hills and take the subsidies on offer; as John Dunning argued, turning them into cash-rich carbon 'prairies' for the privileged by accident, rather than by design as government drifts without a cohesive strategy for farming, food and climate change.

But, if strategy – even the vision thing – eludes the governing class, events are dictating new approaches to food, farming and addressing the climate emergency, in spite of ministerial indifference to anything that purports to be a national strategy for land (Scotland and, to some extent Wales, are treading a different path). Many innovative people working the land are already taking matters into their own hands by embracing nature-friendly farming and enhancing biodiversity – from tenant farmers in Snowdonia regenerating blanket bogs to others pioneering agro-forestry in Cambridgeshire.

Agriculture, after all, accounts for 10% of the UK's greenhouse gas emissions, 'mainly through methane and nitrous oxide from grazing livestock and fertilisers' – although that figure is clearly much higher on some farming estates when the use and abuse of land is factored into the equation, particularly the degradation of moorland.[45]

What's clear, however, is that the debate on returning land to its natural state, whether in the Fens, in the grouse and grazing moors of the hills and uplands and even in the pastoral lowlands, is gaining momentum, under whatever label you want to use.

Up to now, the debate has run the risk of becoming an ideological skirmish between those – sometimes awash with money – pursuing 'rewilding', others preferring a more realistic term of 'wilding', a further quite literal 'grassroots' lobby successfully pursuing 'nature-friendly' and 'regenerative' farming and a network of smaller producers, often organic, who just want to get back to basics: feeding the nation. That means reaching some accommodation, like it or not, to keep farming, at the very least, in part of the Fens – 'the beating heart of British food and farming', says the NFU – between its 3,800 miles of watercourses. This landscape of drained marshes has another distinction: it's the UK's largest fresh produce logistics hub, arguably the centre of its commercial food chain. If it can't be wished away, is there perhaps a pragmatic way forward: balancing rewetting in more vulnerable Fenland areas, while strengthening defences in others?

7

Rewilding: Rich Persons' Plaything or Real Hope for People?

He is richest who is content with the least; for
content is the wealth of nature.[1]

High in the Conwy valley, amid the rugged splendour of Snowdonia, children from a local school are measuring the depth of the peat being restored by far-sighted farmers addressing a new reality: renewing the hills and uplands to help nature's recovery.

For the young pupils from the Ysgol Ysbyty Ifan, the exercise represents science with an edge in the wild beauty of the great Welsh outdoors, beside one of the largest blanket bogs in the principality – the Migneint – where the 11 farmers have grazing rights.

In a riveting film, a nine-year-old well versed in the climate emergency describes how this bog, "vast and remote", stores carbon on a huge scale. "But degraded peatlands damage our environment by releasing greenhouse gases like carbon dioxide into the air," she explains authoritatively in Welsh. "This contributes to global warming."[2]

In this small corner of a splendid National Park, the children are raising an issue at the heart of our challenge to reverse a post-war draining programme meant to 'improve' our moorlands – but which, in reality, only succeeded in degrading them.

To achieve the turnaround on the Migneint, thousands of ditches – opened up to speed drainage decades ago – are being blocked as part of an extensive renewal programme. As the nine-year-old explains, in the helpfully sub-titled film, this is important to "re-wet and restore the Migneint to its natural, wet state … to tackle the climate emergency we are all facing".

Restoring peatlands represents one of our greatest climate challenges, while also presenting cash-generating opportunities for those living near them. Covering an estimated 11,500 square miles, they are prime candidates for being returned to a near-natural state – call it 'wilding' if you must – with

98

one overriding objective: locking carbon in the ground rather than letting it leach into the atmosphere. That's an urgent task because, while only 22% of peatlands are in a relatively good condition, they still account for a quarter of the UK's drinking water, filtered through the peat into streams, rivers, reservoirs – and valued at almost £888 million in 2016.[3] And, just think: they could be delivering more.

These peatlands, then, store enormous wealth: carbon. While the cost of restoring *all* peatlands would be high, it would be considerably outweighed by the environmental payback; initial research shows the estimated 'benefits' for rewetting 55% of peatland alone over the next 100 years, in line with a CCC objective, would be a seemingly eye-watering £45–£51 billion.[4] This opens the way to a new economy: carbon trading, perhaps involving large companies buying credits from restored peatlands – big retailers and transport concerns, for instance. Might this, then, prove a potential saviour for countless upland communities – cash for carbon? – in return for their labours in renewing land where sheep numbers will be drastically reduced?[5]

Across Wales, Scotland and England, projects like the Migneint restoration are renewing landscapes and transforming hills and uplands; none more so than on Britain's largest nature reserve, Mar Lodge, in the Scottish Highlands, on the neighbouring estate of Glenfeshie and on adjoining large holdings, under a collaborative project called 'Cairngorms Connect'. This multi-million-pound programme is labelled 'Britain's biggest habitat restoration project' for a reason: it is awesome in scale, all 230 square miles of it.[6]

Around Britain, there is much to restore. We are paying the price for damage inflicted on these rugged open spaces through draining them to accommodate sheep, grouse shooting and alien tree plantations so dense that any light – and, hence, plant life – is obliterated. Worse, some of these moors are still being burned – to stimulate the young heather favoured by grouse – and overgrazed by sheep, which have wrought damage on our hills and uplands for decades. It's not that we hadn't been warned.

In 1950, for instance, a seminal book by a British botanist noted that overgrazing by sheep was damaging a fragile ecosystem by removing important sphagnum cover. He also cautioned that dense tree planting of alien conifers on the moors was suppressing 'any growth on the forest floor and natural vegetation'.[7] Today, we'd call that damaging the natural environment in a way that contributes to climate change.

The Snowdonia farmers, then, are showing what can be achieved to reverse this destructive trend with collective efforts and formal collaboration through their own Community Interest Company (CIC). With funding from the Welsh Government, the 11 – tenants of the National Trust – are transforming the peatland on the Ysbyty Ifan estate from an unnaturally dry wilderness into a rewetted landscape bursting with new plants and wildlife and worthy of its designation as a Site of Special Scientific Interest. And on the 70,000-acre Mar Lodge estate, 380 miles north in the Cairngorms, promising signs of renewal are similarly emerging.[8]

But how to describe this remarkable rebirth? 'Rewetting', certainly. But in the current vogue, it might be labelled 'rewilding', or simply 'wilding' – although the description is contested by those closely involved and, in truth, has never been formally defined anyway. How about 'land renewal', 'natural regeneration' or 'reinstatement' and 'nature-friendly farming' – for modest numbers of cattle, rather than sheep, still have a place on our hills – or, more likely, 'enrichment'? The latter, as one Snowdonia farmer explains later, certainly encapsulates the social as well as the environmental benefits for both the people living on the land – perhaps an afterthought to some pushing the ultimate 'rewilding' agenda? – and the countryside surrounding and sustaining them.

That leads us into a whole new area: powerful arguments for repopulating rural areas, after years of decline. Should this ideal, gaining momentum in Scotland – 'repeopling' is now entering the vocabulary – not go hand in glove with 'wilding'? Surely the human species, after all, is an important element of nature anyway? The issue is far from academic. It builds on the success of, literally, a ground-breaking resettlement programme in the Highlands and Islands a century ago, when 6,000 new smallholdings were created, by an active British government, for the landless in areas once brutally 'cleared' of families to make way for sheep. Now the case for repopulation is gaining momentum, partly through new Scottish planning legislation, and by a Scottish Land Commission: measures, and an institution which reformers in England can only dream about.[9]

To be fair, Rewilding Britain, a charity pushing the cause of natural renewal, is encouraging a 'balance' between people and the rest of nature 'so that we can thrive together' – a recognition, thankfully, that humans form part of their equation. But words matter. One definition of 'rewilding' at first seems worthy, hardly created for popular consumption: 'the large-scale restoration of ecosystems to the point where nature is allowed to take care of itself ... to reinstate natural processes and where appropriate missing species – allowing them to shape the landscape and the habitats therein.'[10]

The problem with the term – 'a relatively new but contested discourse' – is that definitions seem many and varied. Two academics have tried to make sense of the issue by condensing various descriptions, while acknowledging a popular misconception: 'For some people, the term "rewilding" conjures up images of wolves, wildcats, bears and, to a lesser extent, golden eagles, beavers and wild pigs – but rewilding has a much wider scope ...'[11]

At one end of the spectrum, for instance, they note that the term might embrace regeneration, and reseeding of native plants, shrubs and trees to aid nature's recovery; at the other end, the reintroduction of once-native wild animals, such as beavers, wild boar and lynx, might also help to 'restore natural processes'.

After trawling through a string of definitions from home and abroad, and noting little consensus on a description, the academics provide what at first seems a better description, from the UK's Parliamentary Office of Science and Technology – namely, 'reinstating natural processes that would have occurred in the absence of human activity'" But then, you ponder and think: hang on, surely that 'activity' by people has formed part of a wider transformative process which, for better or for worse, has shaped our landscape since pre-history and should, as the century progresses, be accommodated in a nature-friendly way?[12]

Hype and hope?

Arguably, rewilding, or 'wilding' in whatever guise, has gripped the imagination of some nature lovers – and certainly attracted a government agency dispensing public funds, Natural England – since the publication of a book by Isabella Tree about her and husband Charlie Burrell's transformation of the 3,500-acre Knepp estate, in West Sussex, now a multifaceted business, a brand, offering guided walks, wildlife safaris, accommodation from tree-houses to shepherds' huts and yurts, in addition to other attractions.[13]

Inevitably, given the contested descriptions of 'rewilding', the Knepp enterprise draws both inspiration and scepticism. One senior land manager interviewed, some way north of West Sussex, praised the business acumen of the couple in transforming the estate, with help from Natural England, and substantially enhancing biodiversity. But, after viewing the estate, he added with some passion: "It is very well managed in a nature-friendly way, but it is not rewilding where you leave everything to nature."

A wealth of publicity certainly underpinned Isabella Tree's observation that 'we had no idea how influential and multi-faceted the project would become, attracting policy-makers, farmers, landowners, conservation bodies ... both British and foreign ... a focal point for today's most pressing problems'.[14]

In fact, many transformative projects have emerged around Britain in various guises, under different labels – often quietly, without fanfare – alongside Knepp's celebrity status. They have one overarching theme: placing nature at the heart of regenerating moorland, hillsides, valleys and – yes – farmland. In some areas, outdated and destructive practices, from dependence on fertilizers and pesticides to livestock overstocking – why, on earth, so many sheep? – are finally submitting to nature-friendly farming, restored hedgerows and meadows, cleaner rivers and wilder countryside, as the Lakeland farmer and writer James Rebanks underlines in his book *English Pastoral*.[15] The damaging consequences of previous land-use practices are being acknowledged; significant, reformist strides are being taken. Could this, then, be a quiet, nature-friendly movement in the making, a dawn of 'reinstatement'? Purists might doubtless argue that it doesn't represent their version of rewilding, but something is stirring nevertheless among the restored undergrowth, from North Wales to the Scottish borders, the Highlands and beyond.

That much is recognized by Tony Juniper, chair since 2019 of Natural England, the government agency charged with sustaining and restoring nature. As one of Britain's best-known environmentalists, a former Green Party activist, past executive director with WWF-UK and, previously, head of Friends of the Earth, he acknowledges that England has "excellent targets" for net-zero carbon emissions before 2050 – essential, he says, for nature's recovery, sustainable farming and developing housing and infrastructure. "But the big challenge now is delivery, and the big challenge of delivery is about integrating the capacities of different government departments, and different agencies, to arrive at joined-up outcomes," he volunteers, perhaps betraying mild frustration.[16]

If the state of what we might call the country's natural habitat is a sorry mess – loss of countless species since the 19th century, for instance – Tony Juniper rightly points out that is perhaps hardly surprising, because Britain was the first country to industrialize, with few questions asked at the time about the impact on nature. The consequent degradation of plant and animal life, he notes, was compounded by the "intensification" of agriculture heralded by the Agriculture Act 1947, which, we now know – but our predecessors didn't appreciate – wrought havoc on nature, hedgerows, soil quality and much else. But is the collective mindset now changing in favour of a nature-based recovery? And, if it is, how does that translate into practical action on the ground? Tony Juniper is placing considerable faith in new environmental legislation in England which will herald the creation of 'local nature recovery' panels at county level, building momentum from the bottom up and joining up local initiatives, whatever the dysfunction in Whitehall (my description, not his!).

Whether this momentum amounts to a significant nature-based recovery, or 'wilding' – but let's not quibble with words – Alastair Driver, long-standing conservationist and director of the charity Rewilding Britain, can chart a significant upsurge in interest since 2015. At first, he says, it was a case of contacting people to spread the message of restoring nature. No longer. They now, he volunteers, come to him: "I'm swamped by people contacting me." He talks of being approached by potential and current landowners, with £1 million or £10 million – even £20 million – on hand 'and this is the tip of the iceberg'. Big, public sector landowners, like the Ministry of Defence – with 568,000 acres – are, says Driver, seeking advice about rewilding too.[17]

All that is dwarfed, however, by the ultimate restorative project at Glenfeshie, in the Cairngorms National Park – and, for once, the wilding label seems justified – where Danish billionaire Anders Holch Povlsen is returning 45,000 acres to a biodiverse haven of woodland, incorporating ancient Caledonian pines, intersected by rivers and lochs and encircled by 'a mountain massif the most extensive and wildest of its kind in Britain'.[18]

Alastair Driver, formerly head of conservation at England's EA, speaks passionately about problems ahead, as well as opportunities. "I know we'd be in a horrible, terrible position if we hadn't had these organisations, hundreds of millions of pounds [spent on work], thousands of people like me doing their stuff for decades. But the simple message is, we're still going backwards on climate change mitigation."

To the obvious question – why not ramp up spending? – he has a simple answer: there's not enough money to fund a nature-renewal programme that matches the scale of the climate challenge: "You're talking about trillions," he laments. So other, innovative ways have to be found. Still rewilding, then? "I'm absolutely certain that is the answer ...". But not, he insists, over swathes of the country – just 5%, or under three million acres, will do. Much of that will have to be in Scotland. That prospect certainly divides opinion north of the English border – not so much over the concept of working in harmony with nature, but more the prospect of big enterprises, funded by the wealthy and super-rich, determining land use across the Highlands at the expense of fragile local communities. As the campaigning group Community Land Scotland underlines, over the decades people have had to endure what could be seen as an 'elite view' of 'how the landscape ought to look'.[19]

Nature's recovery

Beyond those 11 enterprising farmers in Snowdonia, a diverse mix of ventures around Britain certainly point towards new beginnings: the 'Wild Ennerdale' collaborative project in the western Lake District, for instance, established in 2003 between the National Trust, Natural England, the Forestry Commission and a water company to restore nature by planting broadleaf trees, reducing sheep numbers and allowing natural regeneration of the valley floor with native cattle. Then there's the extensive tree and shrub planting and regeneration of the 1,650-acre Carrifran Wildwood in the Scottish borders, bought by a local forest trust in 2000 to 'evoke the pristine countryside of 6,000 years ago' by planting thousands of shrubs and trees. More recently, the community trust behind the buy-out of the 5,200-acre Langholm Moor, near the English border, aims to restore peatland, plant native woodland, build new housing and create a small solar and wind-powered energy scheme. Then there's the gem of them all, 190 miles further north: the 72,000 magnificent mountain acres of Mar Lodge in the Cairngorms, owned by the National Trust for Scotland, which, together with the 45,000-acres of the neighbouring Glenfeshie estate (owned by Povlsen) represents either 'rewilding' or 'natural regeneration' – whatever your language – on a truly global scale when combined with the Cairngorms Connect project.[20]

With a reduction in red deer numbers on progressive Highland estates, and a partial reinstatement of the once vast Caledonian pine forest high on the agenda, there's a link between all these projects: a determination to right the wrongs and untold damage of the last century caused by the draining of peatland, more recently the rising population of red deer – kept high for stalking (shooting) – the felling of trees for timber, notably the indigenous Scots pine and, in some cases, the introduction last century of excessively dense forest of alien conifers, suppressing all natural vegetation. This was all in the name of progress – and only a relatively few raised concerns at the time about the planting of these unnaturally dense, green blocks of trees scarring the landscapes of upland Britain, rather than a varied mix of well-spaced, indigenous species.

But, while a post-war agricultural revolution was in full swing during the 1950s, driven by a newly mechanized farming industry creating large field systems showered with chemicals, one leading botanist recorded the regression of a green, pleasant and (in the case of Scotland, the Pennines, Lakeland, and North Wales) one-time wild land. In a seminal book, *Mountains and Moorland*, William Pearsall provided a picture of a Britain which had long disappeared: tree-covered uplands, hills and, 'in early historical times', mountains afforested, such as England's highest (Scafell Pike). He noted that trees were once 'fairly widespread' up to at least 2,000–3,000 ft, while much of our present grassland and moorland 'was probably woodland'.[21]

Subsequently, on moorland, heather would grow long and 'leggy', leaving 'numerous openings' in which other species could develop to create a natural moor of continuous sphagnum and moss cover, alongside the growth of other plants, such as bilberry. But then, through burning, grazing and draining, great, natural, damp and deep areas covered in peat – what today we'd call our most valuable carbon store – were degraded and drained for grouse shooting and for intensive sheep grazing. Today, we're paying a price for that folly – as progressive farmers in Snowdonia, and elsewhere, recognize. But why so much damage? As Pearsall notes, sheep graze 'so much more closely than larger animals that they destroy many more of the smaller (plant) species' – thus 'puddling' the surface, rather than breaking through it, like cattle.

What's to be done? A key report commissioned by the National Trust, the RSPB and the UK's Wildlife Trusts provides a clear path towards renewing hills and uplands – and, crucially, keeping people working the land for nature's recovery – while in Snowdonia farmers are already leading the way. Since 2010, a CIC formed by those 11 farmers, called Fferm Ifan, has reversed the devastating impact of draining what became rough farmland alongside moorland, used for common grazing, the Migneint.

This, then, represents rewetting and renewal, through cooperation and self-help, on a substantial scale – partly addressing the view of Alastair Driver, from Rewilding Britain, that "complementary ways" are needed to match public funding and address global heating. With the support of £696,352 from the Welsh Government's rural development programme, plus other help, the Fferm Ifan project has involved creating extensive woodland, planting and restoring 10,000 metres of hedgerows and blocking another 10,000 metres of peatland ditches – thus allowing the bog to retain more water and carbon. So, rewilding, then? Talk to the Fferm Ifan farmers, and you realize how contested the term has become to them and to others.[22]

I cautiously ask Guto Davies, one of the 11 farmer members of Fferm Ifan, whether 'rewilding' accurately describes their work. The very term grates. He thinks it's an alien concept, a label attached to rich people driven more by personal ambition than by the importance of the human species. His alternative is simple, profound: *"En-rich-ment"*. He emphasizes every syllable with a passion that underlines a commitment to "community, land, language" – in other words, people and place. He can't envisage working individually, "just for myself", for personal gain. Perhaps that explains the donations Fferm Ifan has made to the local primary school and to the nearby community centre over the past few years. They are all in this together.

But, while words are contested, what can't be denied is that, with hindsight, drainage of the 6,000-acre Migneint represented a post-war assault on nature – in the name of raising lots more sheep for their meat – in this rich blanket bog as well as in tens of thousands of similar acres around Britain. Doubtless the consequences were unintended, but the damage was untold to the climate, as well as exacerbating flooding in countless valleys and villages, such as in the Calder valley, described earlier.

In that inspiring film, our young commentator from the local school – Ysgol Ysbyty Ifan – reminds us that, during the 20th century, farmers were encouraged by the government to drain the Migneint, "for keeping more sheep ... to produce more food for the nation ... we now know that drainage has damaged the Migneint as the peat dries and erodes causing carbon dioxide to escape into the air".[23]

She then underlines a sobering statistic: that while only 3% of the world is peatland, it stores more than twice the carbon contained in all the planet's forests – a figure rarely far from the lips of Andrew Baird, Professor of Wetland Science at Leeds University, and a leading authority on the UK's peatland.[24]

We are, let's remember, one of the world's great peat-rich counties, with 'blanket bogs' – a mantle of peat over earth and rock – comprising a fair chunk of those estimated 11,500 square miles of peat cover A pity, thinks Andrew Baird, that bog-land Britain has such "negative connotations"; rather than being considered "wasteland", he says it should be seen more widely as our richest carbon store as well as a first line in defence against flooding.

But he complains that both historically, and culturally, peatlands have been woefully undervalued, ridiculed – think pejorative labels such as 'bog-off', 'bogged-down', 'bog-standard' – and once considered ripe for 'improvement'. Only in the early 2000s, he recalls, were they were finally recognized as a rich carbon store – although still degraded in places today by sheep grazing, commercial forestry, extraction for garden compost and deliberately being set alight in 'controlled' burning to stimulate new heather to attract grouse for game shooting, whatever the consequences for leaching carbon and compounding flooding in communities, like those in the Calder valley.

Therein lies a problem, seemingly easily addressed but bogged down – figuratively – by the inaction of a government unwilling to confront a small but powerful lobby head on: the wealthy, the well-connected and the royals, either owning or apparently gaining some pleasure from shooting grouse over an estimated 550,000 acres of British moorland: an activity inexplicably propped up with millions of pounds in agri-environment subsidies through the EU's CAP.[25]

To be fair, when in charge of Defra, Michael Gove at first warned that a post-Brexit landscape risked "being shaped by vested interests who have either the money, or the connections, or the power".

Private records of meetings between the government and grouse moor owners, released to Friends of the Earth under a Freedom of Information request, show ministers were considering banning moor burning. But Michael Gove advised them instead to sign up to a voluntary agreement 'as it would "help the government demonstrate its intent"'.[26]

What passed for an uneasy peace between the RSPB and the Moorland Association, representing the owners of 144 grouse moors, has now broken down. At the RSPB's 2020 annual meeting, members decided that unless a licensing scheme could be shown within five years to have reduced the environmental harm caused by driven grouse shooting – where beaters drive the grouse towards shooters hiding behind 'butts' – the Society will press for an outright ban. Less intensive, or 'walked-up' shooting, which involves no beating, would be unaffected.[27] As it is, self-regulation by the Moorland Association has neither failed to stop birds of prey, such as hen harriers, being poisoned by gamekeepers, nor always led to improved land management to prevent the flooding risk from burning peatland. Progressive estates committed to regenerating the moors and working in harmony with nature still allow 'walked-up' shooting as well as carefully monitored deer-stalking – particularly in Scotland – to control numbers of the animals.

Food and people

But, of course, wider issues need addressing: namely, where and how 'rewilding' in its various forms – if we want to call it that – addresses the equally pressing issue of keeping people on the land and, crucially, producing the food we need in a nature-friendly way. As Patrick Holden, dairy farmer and leading voice in the organic movement, says, it's not clear where Britain's food would come from if many others followed the example of the Knepp estate owners, Charlie Burrell and Isabella Tree. He is friendly with them. "They're very nice people," he says. "I don't think what they're doing is wrong."

But does this represent the real world, he ponders? And can the groups which Holden calls the 'land sparers' – those who want more intensive crop production on better land so as to release huge tracts for rewilding – reach common cause with the 'land sharers', like Holden, who have moved from chemically intensive farming to growing food and raising livestock in harmony with nature? That's too much of a stretch, thinks Holden. He recalls a chat with an official from Defra on where policy lies on these two fronts. "We'd like a bit of both," came the reply. Holden found this circle impossible to square.

However, a fascinating podcast from the Bristol-based Sustainable Food Trust (SFT) did appear to reveal a vague meeting of minds between Isabella Tree and Patrick Holden, who heads the SFT.[28] In it, Isabella Tree parts company with the leading rewilding advocate, ecologist and writer George Monbiot, apparently disagreeing with his view that gaps in food production caused by rewilding can be met by more intensive farming on huge holdings such as Sir James Dyson's enterprises in Lincolnshire, "I am not of the George Monbiot frame of mind that we can sort this out in a technological way," she said. Isabella Tree's answer is 'regenerative agriculture', with no 'inputs' from fertilizers, and definitely no ploughing, which disturbs carbon in the soil. If this did not amount to a meeting of minds over 'farming in harmony with nature', it seemed to come close to it; the woman who has become a public face of rewilding said she thought they represented "two sides of the same coin". Progress, then?

But will a further intensification of farming, in eastern England and in the Fens – ironically, as noted earlier, the most susceptible to rising water levels and, of course, a huge emitter of carbon – plug the gaps from taking land out of production for rewilding? If Rewilding Britain wants to devote *just* 5% of land to natural recovery, with the majority in Scotland, there seems little to argue over. Surely the bigger issue is the need to turn arable acres, currently used for livestock feed, into land used to grow crops for human consumption – thus taking up some of the slack from the loss of farm land caused by re-wetting part of the Fens? But other forces are at play.

Enter Sir Ian Boyd, former chief scientific adviser at Defra. He wants half of Britain's farmland turned into woodlands and natural habitats to fight the climate crisis. As well as taking arable land out of production his plan would mean the numbers of cattle and sheep falling by 90%. Instead of subsidising livestock production, he thinks farmers should be paid to store carbon dioxide – something the emerging ELMS in England will address – thus helping to prevent floods and providing nature-friendly landscapes for trees, plants and wildlife. "We need a large radical transformation and we need to do it quickly, in the next decade," Boyd has said.[29]

Under his plan to fill production gaps – and this is no longer blue-sky thinking – vast greenhouses would emerge devoted to growing conventional crops through hydroponic or 'vertical farming' – an idea already floated by the CCC, which thinks it offers considerable benefits, with multiple harvests each year under glass. And, of course, Sir James Dyson, in Lincolnshire, has already built a huge greenhouse for indoor production – in addition to other similar enterprises elsewhere, supported by 'green' investment funds.

Could Sir Ian Boyd, then, be ahead of the curve? Others aren't so sure. A coalition of farming interests, for instance, is already arguing that the UK government's ELMS is woefully thin on measures to sustain and increase domestic food production; some conservation groups, on the other hand, such as Rewilding Britain, have given ELMS a cautious welcome. Boyd argues that farmers are sitting on a potential "gold mine" from likely payments for renewing degraded land and – who knows? – from carbon trading too. That, at least, is a welcome acknowledgement that – as Patrick Holden says – the human species represents an important element in a nature-based approach to land use. And the importance of people in the great outdoors is an issue at the heart of the landownership debate in Scotland – and the place 'rewilding' (or whatever term we care to use) occupies within that.

Repeopling, renewal and ownership

Concentration of land ownership has a direct influence on the public interest with potential adverse consequences through the exercise of market and social power ... amplified by large-scale ownership.[30]

You might think that such a bold statement represents the blindingly obvious in a country where 432 landowners account for half of all privately owned land; where several billionaires own swathes of its most varied, rugged, mountainous and valuable acres bursting with wildlife, native forests and

**7.1: Remnants of Caledonian pine forest,
Glen Derry, Cairngorms**
Source: Murdo MacLeod

**7.2: Seaside spectacular:
Ardroil beach, Uig, in the Outer Hebrides**
Source: Murdo MacLeod

7.3: Cairngorm spectacular: Loch Avon. Swathes of the Cairngorms – now a national park – are now being regenerated, or 'wilded'
Source: Murdo MacLeod

7.4: New wetland is both attractive and nature friendly

7.5: Renewal in the fields of the National Trust's Wallington estate in Northumberland

7.6: Talisker, Skye: an area settled by crofters 100 years ago under radical land reform legislation still to be equalled in Britain
Source: Murdo MacLeod

rare, indigenous plants; and where the old aristocracy and the royal family still wield considerable power on estates rich in some of the pleasures they most value: hunting, shooting and fishing.[31]

This, then, is Scotland. The divergence from its southern neighbour, in this fragile union of Britain, is stark both in policy and in topography. Here liquid assets – the rivers, lochs, wetlands – and the surrounding mountains are literally worth millions. And the state, in the form of the Scottish Government, knows it wants change. But here's the question: how to achieve an ultimate goal of an equitable distribution of land – in a country still with 'the most concentrated pattern of private land ownership in the developed world'[32] – beyond a series of reform Acts which have, in varying degrees, triggered community ownership of around 520,000 acres, mainly in the Highlands and Islands, and more recently in the lowlands near England? For some in Scotland, this legislation appears modest because it fails to tackle one overarching issue: that, in spite of land reform, little so far has halted the concentration of landownership[33] and the continuing sale of substantial tracts to those with, sometimes, little interest in either the environment or local communities.[34]

But from a wider, British perspective, the measures appear ground breaking. Bluntly, as things stand, they could never happen in England, where proposed planning deregulation will further undermine the abilities of communities to determine their future in a country where few questions are asked in officialdom about either the ownership of land or its use and abuse. But, as noted earlier, a Westminster government – well before Scottish and Welsh devolution – once had the determination to transform landownership and create the conditions for land settlement through substantial state intervention in England, Scotland and pre-partition Ireland.

Consider just one significant aspect of land reform, which should – but undoubtedly won't – provide a guiding light for Westminster: the creation of a Scottish Land Commission, an agency of the Edinburgh government, in 2017. It is charged, in its owns words, with 'working to create a Scotland where land is owned and used in ways that are fair, responsible, and productive ... where people are able to shape and benefit from decisions about land'. Importantly, the Commission, based in the Highlands capital of Inverness, is not afraid to raise its voice and challenge the status quo, not least of pointing to 'convincing evidence' that highly concentrated landownership can have a detrimental effect on rural development.

In other ways, Wales benefits from coordination through a government agency, Natural Resources Wales (NRW) – combining land use, environmental protection and forestry, for instance – which proved particularly helpful in supporting those 11 farmers in north-east Snowdonia. Significantly, NRW has launched a task force to spearhead a post-COVID-19 'green recovery', and develop ideas linking climate action with job creation, economic growth 'and other development priorities.'[35]

But, alas, in England there's little momentum for reform – yet. The case for a reborn English countryside agency, both advisory and with some delivery powers, similarly semi-independent of government – and we did have one earlier this century – becomes stronger by the month. But, as things stand, England has neither a cohesive rural policy, nor – importantly – does a UK government have an overarching food policy for this island built around resilience.

The Scottish Land Commission performs one vital role: it ensures that land reform remains a live issue, and cannot be side-lined – in a way that any discussion about land in England, and the management of the countryside, is, frankly, often marginalized, while land reform itself is a fringe issue and definitely off the agenda. A pre-1945 commitment from Labour to work towards the public ownership of land, with the Town and Country Planning Act 1947 'nationalizing' development rights seen as a first step, is now barely remembered. Effective planning is no more in England. Scotland, on the other hand, has its third statutory land-use strategy, for the five-year 2021–26 period: 'Getting the Best from Our Land'. Wales, similarly, has a national plan, alongside specific well-being legislation and even a 'future generations commissioner', with powers over public bodies. England, again, is the outlier.

Significantly, in Scotland, momentum behind societal control over that most basic resource – our land – is now driving change in the lowlands. Close to the English border, in the small town of Langholm, a community trust raised sufficient funds in 2020 to buy 5,200-acres of surrounding moors and hills from Buccleuch Estates, until recently the country's largest private landowner, for £3.8 million. They aim to develop what's known as Langholm Moor as a nature reserve, promoting tourism while planting native woodland and restoring peatland; in short, a form of rewilding. The trust, Langholm Initiative, talks boldly of placing nature restoration – and regeneration of their town – ahead of profit.[36]

But this latest community land buy-out tells us something else about the changing pattern of Scottish landownership. Buccleuch Estates, owned by trusts created for the current duke – Richard Scott, 10th duke of Buccleuch – and his family, is now shedding land; it now has *only* 186,000 acres left in Scotland (with a further 11,000 on the Boughton estate in Northamptonshire and a grand house, described as 'the English Versailles').

Others are taking up the slack; the biggest private landowner in Scotland – once Buccleuch – is now a Danish billionaire. Anders Holch Povlsen, with substantial retail interests in Denmark and in the UK – not least as the biggest shareholder in the online clothing store Asos – has steadily built up a landholding of 220,000 acres embracing 12 Highland estates spread mainly across the Cairngorms and Sutherland, at the northern tip of Scotland.

Here, then, are two contrasts. Richard Scott has already made clear his frustration and – yes – anger at the pace of Scottish land reform. He has spoken in near-apocalyptic tones about a "pivotal moment" being reached with his three main Scottish estates shrinking by a quarter: "What is the point of hanging onto it all and being shot at by the Left for your pains?"[37] No surprise, then, that the landholding of Buccleuch Estates is shrinking under the executive chairmanship of Benny Higgins, former Tesco Bank chief executive. Since 2010 Buccleuch Estates has sold, or agreed to sell more than 41,000 acres – and Higgins has said that, perhaps, another 50,000 or more are likely to go. He would like the company to be known for its diversity of interests, which include property development, rather than keeping the label of 'one of Scotland's biggest landowners'.[38]

By contrast, since 2006 Anders Povlsen has been building up his 220,000-acre land holding – including the 42,000-acre Glenfeshie estate, bought for £8 million in 2006. As an essential part of the ambitious Cairngorms Connect project – billed as Britain's biggest 'habitat restoration' programme – Glenfeshie represents the ultimate in rewilding for many enthusiasts.

Significantly, the Povlsens, through their operating company in the Highlands – Wildland – say they are "custodians" of the land, and speak of a "deep connection with this magnificent landscape". By investing £50 million in rewilding projects, they have pledged to restore their extensive holdings to "their former magnificent natural state and repair the harm man has inflicted on them": tracts of new and improved native woodland, reborn peatlands, and even the reintroduction of native species, such as lynx, are all on the agenda.

Tim Kirkwood, chief executive of Wildland, insists plans for the Povlsens' estates are underpinned by a long-term commitment to enhancing landscape and ecology: "I cannot stress enough this is not land ownership for the sake of land ownership."[39] The Povlsens' commitment to Glenfeshie, under a progressive land manager, is endorsed by Roseanna Cunningham, the Scottish Government's Cabinet Secretary for the Environment, Climate Change and Land Reform until elections in May 2021. She describes the family as "conservation landowners" with a deep commitment to the Highlands.[40] Alongside Glenfeshie, however, the nature recovery on the 72,000-acre Mar Lodge estate, owned by the National Trust for Scotland (NTS), is equally impressive – and highlighted by Rewilding Britain as one of the most important areas for nature conservation in the British isles: 'wild beauty unfolds around the estate, creating a truly astonishing landscape'. Even that barely does justice to an estate with 15 munros (mountains over 3,000 ft high) and four of the five highest peaks in the UK, including Ben Macdui, the second highest; an arctic alpine plateau, rich in plant life; and – like Glenfeshie – an expanding population of wildlife: red squirrels, golden eagles, capercaillie, black grouse, pine martens and much more.[41] And, as on progressive holdings elsewhere, controlling the red deer population – kept artificially high on 'sporting' estates through selective culling to protect the breeding hind population – is one key to regenerating woodlands, shrubs and plant life.

Rebirth of native plants and species, then, begins with culling large numbers of deer – vital for the renewal and rebirth of an ancient Caledonian pine forest – and, more recently, with ending 'driven' grouse shooting to protect and enhance invaluable peatland. In his ten years as manager of Mar Lodge – which celebrated 25 years under NTS ownership in 2020 – David Frew has seen the deer population fall from almost 5,000 to 1,600, as well as "startling increases" in plant and natural animal life. Where deer kept trees and plants dormant, destroying saplings – "there was nothing above deer height", says Frew – trees are now "bursting through" the heather; on higher-altitude montane areas, birch, juniper and scrub are emerging. Frew confidently expects, in a new census, to see that their efforts have led to around 4,000 acres of new woodland emerging. But would he describe this turnaround as 'rewilding'? "It is not something I am very keen on," he responds. "It's very easy to misinterpret." So what, then? He settles on "natural regeneration", and leaves it at that.

More broadly, however, 'wilding' – let's briefly stick with the term – raises profound questions, highlighted by the farmers in the Fferm Ifan CIC in Snowdonia: how do people, and the places they inhabit, coexist alongside a nature-led recovery? Bluntly, when does the rock of highly concentrated landownership collide with the hard place of community interest and ambition? If Rewilding Britain wants to turn 5% of land over to nature by 2100 – say, around three million acres – then, as its director Alastair Driver acknowledges, the Highlands and the islands of Scotland will be critical in achieving that goal, with perhaps 3% of that total. The human species, then, must surely form one vital element of that recovery – and, as Rob Gibson, former convenor (chair) of the Scottish Parliament's rural affairs committee points out, any rational, historical analysis of land use shows that people and nature have long coexisted and always worked well in harmony.

To be fair, Anders Povlsen's Aviemore-based company stresses their custodianship and community focus as job creators – the estates have fine, boutique residences for rental – and thus, unlike others, they truly value the great Highland outdoors. Hence, the argument goes – and this was put to me by, arguably, the leading advocate of rewilding in the UK – who else would have the money, the commitment and the long-term vision of a 200-year-long project to gradually return Glenfeshie to its natural state, with even the introduction of once-native mammals a possibility?

But does Wildland's vision conflict with the ambition of fragile communities in, say, Sutherland and beyond, who can see the potential for expanding affordable, rented housing for essential but relatively low-paid workers in – say – the caring, health, education, forestry and land management sectors? In short, where do people figure in the ultimate 'rewilding' equation? While Povlsen's transformation of Glenfeshie – the first of the Highland estates he acquired in 2006 – has been described by a neighbour as an "absolute ecological delight", he has attracted some opposition in Sutherland, in the far north-west. In early 2021 the chair of a community estate in the township of Welness – gifted to local crofters in 1995 by an absentee landlord – was scornful of a "rich billionaire trying to impose his vision on the north coast of Scotland ... our estate is a wild estate ... we do not need [someone] to re-wild it."[42] A year earlier, the local daily newspaper reported that a couple in the Sutherland village of Tongue, apparently outbid by Povlsen's Wildland company for a former bank building, are 'part of a growing groundswell of local concerns that the Danish tycoon's deep pockets are creating unfair competition'.[43]

As it stands, rapidly rising house prices in the Highlands and islands and in rural Britain – partly spurred by a post-COVID-19 demand for home working, and for second homes – risk undermining attempts to both modestly repopulate the countryside and provide homes to keep young people in rural jobs, and attract others to the countryside. While opposition to Anders Povlsen can, possibly, be overstated – and he probably has critics, and sceptics, in the highest establishment places as well as in some communities – wider issues are clearly at stake here.

Unlike deregulatory England, the Edinburgh government sees a new Scottish Planning Act as a way of increasing the population of rural areas. Might this legislation be used to increase the supply of affordable homes in places desperate for low-cost property in a country where, unlike England, the right of sitting tenants to buy social housing stock has been abolished? Rob Gibson, former chair of the Scottish Parliament's rural affairs committee, and still active in land campaigning, certainly hopes so.

Learning from history?

But much as the drive for community ownership in Scotland exposes the absence of any modest reforms of land use in England, the current measures – radical as they might seem – still appear a pale shadow of what emerged in the Highlands after the First World War. The Land Settlement (Scotland) Act 1919 amounted to a quiet revolution, 'arguably the most transformative land reform Scotland has seen'.[44] Significantly, the anniversary was marked by a commemorative booklet, produced by the Scottish Land Commission. Its title certainly caught the popular mood: *Repeopling Emptied Places.*

Viewed in today's context, what's remarkable is the speed with which the Westminster government – then running Scotland – moved to resettle the landless in areas emptied of people to make way for sheep in the infamous Highland Clearances in the mid-18th to 19th centuries. And the legacy, indeed, is enduring, as the historian, academic and Highland activist Jim Hunter recalls.[45] In the summer of 1923, for instance, in the North Talisker area on Skye alone, he records 'family after family moving to 68 crofts taking the place of a sheep farm – part of a 60,000 acre estate bought by the then Board of Agriculture for Scotland'. Overall, the land settlement legislation – which, interestingly, has never been repealed – gave the British state compulsory purchase powers which resulted in the government creating 6,000 new smallholdings (or crofts) around Scotland "in the teeth of landowning opposition and demands for compensation", according to the former Scottish Labour MP and UK Energy Minister, Brian Wilson, who lives on the island of Lewis. Yet today, he rails, "there is minimal political challenge to the right of private landowners to act as masters of their great fiefdoms".[46]

Wilson, a long-time opponent of the Scottish National Party (SNP), sees no reason why the 1919 legislation, still on the statute book, should not form the part of a post-COVID-19 recovery plan for Scotland – while the campaigning group Community Land Scotland wants measures to create a 'public interest test' for prospective buyers to pass before being allowed to buy land. And they certainly have many sympathetic ears in the Scottish Parliament.

When the Land Reform (Scotland) Bill finally passed through that Parliament in March 2016, it was greeted in Holyrood by plaudits ranging from "new dawn for land reform" to "transformative". The legislation included new protections for tenant farmers, and an end to tax relief on sporting estates – for 30 years they didn't pay business rates – accompanied by a Scottish Land Fund with a modest £10 million available to help community land buy-outs. There was one omission, however: the inclusion of amendments which would not only restrict the amount of land an individual can own, but also prevent landownership through offshore tax havens, which are used as safe places to register ownership in family trusts. These were voted down by SNP and Conservative MSPs (Members of the Scottish Parliament). One SNP dissident lamented that her government had "passed up the opportunity to deliver real reforms".[47]

Acknowledging concern over the concentration of ownership, Roseanna Cunningham, the outgoing senior land reform minister, describes land reform as a "process not an event". She is convinced further reformist moves will follow, but cautioned that the European Convention of Human Rights placed restrictions on what amounts to state land-grabs. "Landowners have rights too," she stressed. But, while complementing Anders Povlsen on his ambitious rewilding project, describing the Dane as a "conservation landowner", she added that "it is not in the gift of one individual to make a wholesale rewilding decision ... people cannot rewild willy-nilly ... putting a big fence around is not rewilding".[48] Which, to be fair, is not what Povlsen can do, even if he wanted; Scotland, after all, has the cherished freedoms of 'right to roam'.

What next?

Perhaps it's time for a reality check and to put 'rewilding', or whatever term we care to use, into perspective. While we should recognize that transforming places like Glenfeshie, Mar Lodge and other estates in the Cairngorm Connect project with ambitious nature-led recovery programmes is generally welcome, there's an equally important issue to address as well: 'repeopling' rural Scotland, and Britain, where appropriate. It has been achieved before – as the 1919 Act has shown in Scotland, alongside other ambitious moves in England, which created not only county council farms but also the LSA and, hence, hundreds of smallholdings. Any action – back to the future? – could easily take a cue from initiatives in relatively recent history when an active state compulsorily bought land at 'existing use' value – rather than at inflated development prices – to create many of the UK 's 34 post-war new towns and ensure sufficient houses were built for a population desperate for new homes. And, in Scotland at least, those pursuing 'repeopling' think similar measures are now needed to provide homes in fragile rural communities; time, thinks Rob Gibson, long-standing land reform campaigner and former MSP, to test the system and step in to acquire land for housing at reasonable 'existing use' (rather than development) value to help renew communities. Significantly, the Scottish Land Commission has now thrown its support behind reformist moves for affordable housing. In England, by contrast, fears are growing that planning 'reforms' being pursued by the Westminster government will further reduce the supply of affordable homes and, hence, undermine further those communities in need of affordable housing.

People and nature

None of this is to minimize the action needed for a nature-led recovery. And here's the challenge: as Alastair Driver from Rewilding Britain underlines, if the billions of pounds necessary to renew and rework our countryside are simply beyond the reach of governments, how can global institutions be mobilized – angel investors, pension funds? – to work with countries and communities to renew our land and the settlements therein? And how do we put a price on nature anyway, beyond – in Britain at least – a fevered market in land speculation fuelled by tax breaks and, sometimes, by land agents who put monetary gain for clients ahead of nature's renewal, soil quality and – yes – public interest and, hence, truly sustainable management of our most basic resource?

Does the Cairngorms Connect initiative, which can safely call itself 'Britain's biggest habitat restoration project', covering 230 square miles, point one way forward? Partly backed by Arcadia, a £23-million fund established by billionaire philanthropists Peter Baldwin and Lisbet Rausing, it is effectively a public–charitable partnership partly aimed at restoring peatland and joining up areas of an ancient Caledonian pine forest with improved land management and a much-reduced red deer population. Already they are talking of 'clear evidence of young woodlands spreading, peatlands and bogs showing signs of repair, and the early recovery of rivers moving more freely across flood plains'.[49]

Perhaps another way forward might involve those blessed with money, and land, to at least – under legislative pressure, if necessary – provide modest acres at 'existing use value' for housing, smaller-scale farming and community benefit, in the interests of reworking our countryside for the benefit of all.

Just think what happens when rural communities mobilize. In Snowdonia, for instance, a low-carbon future, supported by NRW, is the driving force for the 11 tenant farmers on the Ysbyty Ifan estate – although rewilding is not a description they recognize. The holding came into the care of the National Trust after being transferred to them by the UK Treasury in 1951, which received it in lieu of death duties from the estates of a late aristocrat – and it's proving a showcase for a Trust keen to demonstrate its leadership role in reaching its own net-zero carbon target by 2030. "We can imagine, inspirationally, large tracts of land where nature is definitely the priority outcome – more trees, rougher ground, still some animals on the landscape ... but working with nature," enthuses Patrick Begg, the Trust's director of natural resources.

For his part, Begg sees less of an ideological divide now between those favouring rewilding and others attracted to – say – nature-friendly farming. "Three to five years ago it seemed there were only two alternatives – totally wild or utterly farmed – and that was producing some tensions but more recently ... this sounds terribly mealy-mouthed ... there's more of an accommodation ... accepting other views ... a place for all of us," he volunteers.

In all this, we can easily be deluded into thinking that change is afoot; that Britain is finally stirring under the weight of both the climate emergency and the unequal distribution of land. But too often money, or rather, 'investment' – a safe, tax-efficient place to dump cash, with few questions asked – determines the use (and, sometimes, abuse) of that most basic resource: our land.

I'm often amazed that, in spite of the injustices wrought by the Enclosures in England – in reality, a land-grab of an estimated six million acres of common land by a dominant aristocracy – there's been little sign that this appropriation spree, ending in the later 19th century, fuelled reformist moves to compare with the appetite for change in Scotland driven by the brutality of the Highland Clearances 200 years ago.[50]

We sometimes need reminding that the enclosure of common land deprived thousands of rural workers of their collective livelihoods – to the extent that, by the 1870s, the idea of collective land rights in Britain had been largely extinguished.[51] We need to bear this is mind as an emerging 'rewilding' agenda gains support among some in England's Defra; communities, and people living on the land (as opposed to owning it), must be at the heart of the substantial changes envisaged in English land use.

This is important. The new ELMS regime – paying farmers and landowners to improve natural habitats – will signal both a physical and a philosophical change in land use, with little public discourse. Significantly, while ELMS has been given a cautious welcome by the Rewilding Britain group – so far, so good, thinks Alastair Driver – farmers' organizations fear the regime risks prioritizing the environment at the expense of food production. Might they have a point? As Patrick Holden, doyen of the organic food movement and West Wales dairy farmer, cautions: "Although we do want our farms to be a bit on the wild side, I don't think we can take huge areas of UK land out of production either to reforest or rewild it … there won't be enough land left to feed everybody."

As recorded earlier, the organic and the ecological farming movements believe nature-friendly farming, and measures to enhance biodiversity, can happily coexist. But Holden's view is that rewilding could mean "shutting the gates, leaving everything completely alone". What, asks Holden, about communities themselves? "People are part of nature and that's something we should not forget."

Unless we're careful, our concentrated structure of landownership throughout Britain, and a new wave of landowners awash with cash, will determine the land-renewal strategies they perceive, at best, to be in the national interest, buoyed with ELMS subsidies and given a cloak of respectability through community 'consultation'; insecure communities, who should be part of the solution and, thus, gaining any rewards for enhancing nature, will then have to fall into line. In England, a substantial chunk of our land is already in the hands of the aristocracy and the new rich: around 36,000 people, 0.6% of the population, own half the rural land alone.[52] And in Scotland, whatever the grandstanding over land reform, a combination of private owners and the old aristocracy, often with competing aims, call the shots – sometimes, quite literally – on land holdings still dubbed 'sporting estates'.

We are, in short, facing the biggest upheaval in land use in over 70 years, with little debate. Largely under the radar, a new revolution is unfolding with little public discourse – yet its consequences could be profound. As things stand, farming groups – the NFU, the TFA, for instance – fear that enhancing the natural environment, so-called 'public money for public goods' – important as that is – seems to be taking precedence over food production. Whether the two can – should – co-exist is a debate becoming more intense between a coalition of farming interests on the one hand and the Westminster government on the other, unsure how to proceed in the brave new world outside the EU's CAP – with Scotland, and Wales, taking a lead on multiple fronts.

As the former chief executive of the (English) EA, Baroness (Barbara) Young has argued, Westminster could well learn from the Scottish and Welsh approaches to integrated land use to design a "bespoke" English framework addressing "competing and often conflicting" government policies.[53] And among them is the case for renewing country towns and villages.

8

Communities Renewed
or Housing Denied?

'My people will abide in peaceful habitation in secure dwellings...'.[1]

To renew our land, and rework the countryside, we need thriving villages and small towns: at their best, inclusive places with schools, shops, a post office, ideally a health centre-cum-neighbourhood hub and, of course, a pub. Most of all we need affordable, secure homes for those on low and middle incomes who underpin communities. Think of health and social care staff, shop assistants, those creating local food networks, land managers and farm hands – in greater demand to replace departing EU workers – and teachers, for a start. To achieve all this, as Scotland and Wales demonstrate, we need above all a functioning planning system at the heart of local democracy to assess and deliver community needs – none more important than affordable housing – and meet aspirations.

Behind the enduring images of timeless villages with period homes around manicured village greens – and of more remote spots offering solitude and spectacular scenery by the mountains and the sea – lies a hidden crisis. Rural Britain has, in large part, become the preserve of a moneyed elite, abandoning larger cities – an exodus intensified by the COVID-19 pandemic – while younger people, born in the countryside, are travelling in the opposite direction, often reluctantly, because they can't find affordable homes.

This is economically illogical. It represents a failure by successive governments to truly value the foundational 'worth' of people to communities in a country – deregulating England, in this case – more attuned to asset wealth and perceptions of status, than to strengthening the base on which to build houses, communities and a good rural society – in short, valuing the low-paid people, in jobs we take for granted, who kept the country running during a year and more of a pandemic. They are vital to reworking the land to benefit us all.

Where are these essential workers and the young aspirants keen to remain and work in their communities – add value to them – expected to live as new jobs emerge? Take one area: addressing the climate emergency by adopting nature-friendly farming, and progressive land management – rewetting and restoring the land, from regenerating blanket bogs to cultivating non-dense native forests, to replanting hedgerows and, hence, encouraging wildlife. That comes with a price and should deliver an income, admittedly yet to be fully calculated, for (who knows?) a new breed of land managers and essential workers.

The possibilities, which Natural Resources Wales (NRW) is addressing with a task force to accelerate a 'green' recovery – and create jobs – seem endless. In England, the Westminster government needs to ensure that environmental payments in the emerging ELMS regime don't just end up in the pockets of the big landowners (and City types smelling money from carbon trading) and are, instead, spread more equitably across the landscape. Surely community regeneration must go hand in glove with nature's renewal; people working in harmony with land and landscapes.

The potential is there, if we seize the opportunities outside the EU after escaping the straitjacket of its CAP, to renew our land and, literally, rework the countryside.

As it is, a drift from the land might be inevitable as farmers, often approaching retirement age, take advantage of a Defra exit scheme in England – Scotland and Wales are developing other plans – due in 2022. It will offer a lump sum payment, capped at £100,000 and aggregating farming grants up to 2027 (the average farmer has been getting £20,000 subsidies annually under the old EU system). While some farmers own their own homes, many tenant farmers might not. On some estimates from the Tenant Farmers' Association, up to 1,000 might leave the land. And the social implications of what amounts to self-redundancy – where will the departing farmers live? – have yet to be quantified.

As we'll see in one extensive Outer Hebridean community enterprise, successful rural economies depend on one key ingredient, above all else: affordable local housing and, crucially, access to land on which to build homes. And the North Harris community trust has that in abundance: 64,000 magnificent acres of mountain, moor and spectacular beaches, bordered by the sea on three sides.

Others aren't so lucky. Without relatively low-cost housing, and often denied access to land and, hence, to finance, the omens have not been good for many, although in parts of England some places – against the odds – are managing to provide affordable homes, although not in sufficient quantity to stem the drift from the countryside. All this begs a leading question, once answered by reforming governments: whether to buy land at 'existing use' (or agricultural) value – compulsorily, if necessary – for badly needed housing and community renewal ventures. This is being pursued by land reformers in Scotland and, notably, by a former chair of the Scottish Parliament's Rural Affairs, Climate Change and Environment Committee. Rob Gibson sees the concept gaining traction – drawing inspiration from the Land Settlement (Scotland) Act 1919, which created 6,000 crofts (smallholdings), and is seen as the most radical land reform legislation ever undertaken in Britain.[2]

Time for action

Certainly, radical measures are overdue. With a steady drift of the better-off from cities to the shires, we're seeing the polar opposite among the less well-off struggling in the countryside: rural flight. This has been given sharp focus in research showing that, since 1981, Britain's villages and country towns have lost more than a million people under 25 – but, at the same time, gained more than two million aged over 65, attracted to a rural retirement idyll in a green and generally pleasant countryside.[3] Since then, the COVID-19 pandemic has added a new dimension to urban flight: the ideal among younger workers blessed with decent incomes of home working. Could this prove to be the new normal in village, country town and shire, with all the implications for house prices? Estimates show that four in ten potential buyers seeking homes are now considering fleeing cities.[4]

As rural prices are pushed up further, pity those essential workers providing the facilities – from healthcare to retail, hospitality and teaching – demanded by the new rural classes. A survey in 2020 underlining 'demographic shifts changing the face of rural communities' highlighted the shortage of affordable housing and the consequent 'volatility' of rural economies, vulnerable to insecure and low-paid employment.[5] It spoke of this 'new normal' post-COVID-19, with a trend of migrating city dwellers putting so much pressure on rural housing markets, and jacking up prices that, on a 'conservative estimate', 124,000 new rural homes will be needed by 2030 to meet demand. That means carefully planned places, meeting the need for both affordable and full-cost homes in specific market towns and villages, rather than estates, plonked randomly in the countryside and leading to 'rural sprawl'. But the latter will be the likely consequence of emerging centrally driven, deregulatory and developer-led planning reforms in England. Environmental groups, as well as Conservative councils, are horrified.[6] How, they might sensibly ask, does this car-dependent, carbon-intensive free-for-all align with Defra's new ELMS regime built on restoring nature in the countryside and encouraging biodiversity?

Where to live?

Amid this confusion, you might also wonder how essential workers survive in the countryside, with affordable accommodation in desperately short supply and social housing, sadly, a historical relic – a legacy of the late Lady (Margaret) Thatcher – now that so much of it has been sold off. In one Hampshire village, for instance, the volunteer running a small, local housing association, speaks despairingly of care workers being forced to share houses, and of local teachers unable to afford homes in the villages where they work, while, in the next breath, recalling the unprecedented pressure from city dwellers, with spare cash, seeking a new life in the country. "The demand [for houses] in our village [Grayshott, in Hampshire] is relentless; on the market one minute, snapped up the next," sighs Teresa Jamieson.

Against this background, the supply of new social homes in England has fallen from a steady, if inadequate, flow to barely a trickle – more than halving in five years to only 1,090 annually, according to government figures.[7] Assessing supply, however, is a game of smoke and mirrors, because the government's definition of 'affordable' is 80% of a full market rent – often beyond those on the minimum wage, on zero-hours contracts or in seasonal work. For them, there's little on offer: a dilemma underlined by an illuminating documentary which showed that from 2011–12 to 2017–18 alone the supply of new social housing in England had fallen by a staggering 83%.[8]

What to do? Aside from the obvious response – build more social homes rather than concentrating entirely on promoting and subsidizing home ownership and other forms of 'affordability' – the answer partly lies with the valiant efforts of countless public-spirited people striving to run either local community land trusts or even small housing associations: all dedicated to creating inclusive places for all, rather than for a privileged few. But it's clear that something extra will be needed from a UK government running England, seemingly bent on ideologically driven planning reforms which will only exacerbate the crisis by loosening rules governing the amount of social housing in developments – rather than providing practical solutions to provide genuinely affordable homes. Meanwhile, Scotland charts a more progressive course, distinctly different from Westminster, having abolished a long-standing legacy of the late Margaret Thatcher: the right to buy public housing stock (once called council housing) at discounted prices in an unprecedented cut-price sale of public assets. Since the early 1980s, around two million social homes have been hived off in Britain, several hundred thousand of them in rural areas.[9]

Yet, amid the indifference in government to sustaining and increasing genuinely affordable housing in the English countryside, some communities are shining a light on how local action, with pump-priming from cash-starved local government, can prove transformative.

Turnaround towns

For countless years I've been travelling fairly regularly to Scotland, along the winding A697 through the small border town of Wooler. It is neither well to do nor down at heel but, rather, a traditional working town which has neatly repositioned itself as the capital of the Cheviots, partly capitalizing on tourism and the great outdoors of the Northumberland National Park on its doorstep. But while its location is stunning, Wooler could easily have followed countless other places in a familiar spiral of decline; seemingly left behind, ignored by decision makers, starved of essential services, as big cities powered ahead (the town is equidistant between Newcastle and Edinburgh).

It's a familiar story throughout Britain. Banks close – two closed fairly recently in Wooler. Local industries contract. Shops shut. Libraries disappear. Bus services are axed – and, slowly, the life blood of a community is sucked away.

Yet places, like Wooler, have defied the odds through local endeavour. Wooler will even soon have a new distillery: Ad Gefrin – 'by the hill of goats' (named after wild goats in the nearby hills). From modest beginnings in 1996, a local community land trust, or CLT, has grown into a significant economic player. It has an annual revenue of over £100,000, capital assets of £3 million, and operates, among other things, a large community hub – the Cheviot Centre – which includes advice and accommodation for start-up businesses, a county council library, a branch of Newcastle Building Society, a police 'station', meeting rooms for activities drawing all ages and much else.

In addition, the Glendale Gateway Trust, as it's known, owns five high-street shops and has bought a vacant bank building. It has a youth hostel, has delivered 20 affordable homes, and – who knows? – might create more of them. "What we've done is perfectly replicable," insists Tom Johnston, outgoing chief executive of the trust, a member of a growing national CLT network, which stretches from Northumberland to Cornwall, with over 300 member trusts. "We sat here and thought 'there's a risk in doing this'," recalls Tom, a former colliery tradesman. "But we took the risk and it paid off." He acknowledges they are sometimes accused of being "too property-based", but stresses they remain "keen to do more on the high street". Above all, he says, they've "galvanised the community ... sent a clear message that if there are things to be done we can get on and do it".

Almost 500 miles south, at St Minver in Cornwall, another trust was born out of different circumstances: desperation among locals for affordable housing in an area swamped by second homes and holiday lets. The

St Minver CLT, formed in 2008, was the inspiration of a late builder and parish councillor, appalled that his own workforce couldn't afford to live in their home parish, "and he wanted to do something about it", recalls a trust member.[10]

An initial grant of £5,000 from the former North Cornwall District Council was followed by a secured interest-free loan of £544,000. Land was acquired from a local farmer – similarly appalled that locals were priced out of local housing – who then paid for the detailed design of 12 three- and two-bed bungalows. These were built entirely by the new residents, some of them tradesmen; others picked up skills and the rest acted as labourers. The loan was repaid when mortgages were granted to the new owners, with one stipulation: if a house subsequently came onto the market, the CLT would have first refusal either to buy it or to nominate a purchaser to ensure that the property stayed with the local community in an area where second homes dominate the local property markets. A further eight self-build homes have since been constructed by the residents, along with four affordable rented properties for a local housing association. Not surprisingly, the enthusiasm of the self-builders is palpable, with one view paramount: "We are not only building houses for ourselves, but for our neighbours – and, if we can do it, others elsewhere can."

But demand for affordable homes, not surprisingly, dramatically exceeds supply in a county where prices are going through the roof. Figures compiled by the Halifax mortgage lender show that the Cornish village of St Mawes, for instance, had the biggest increase in average prices of any British seaside town in the 2020–21 year – soaring 48% to £502,000.[11]

Amid a hectic market in land trading, driven by the urban flight and the demand for second homes, CLTs highlight one overarching issue: an acute shortage of social and affordable homes, marginalizing a whole (younger) generation – "and a lot of that is down to the land market", according to Catherine Harrington, former director of the CLT Network.[12]

Land is sometimes gifted to CLTs on condition that it is used only for affordable homes – either owner occupied or rented. Typically, the CLT keeps the freehold, putting a resale condition on a property to retain its affordability and ensure that it is not subjected to the excesses of the open market. As Harrington has said, governments "have ignored the land question for too long". Actually, at one time, they didn't.

In the past, governments used their clout to create thousands of smallholdings in Scotland and in England after 1918, the former still providing an enduring legacy, the latter wound up in the early 1980s. In Letchworth Garden City, Hertfordshire, surely a model for the CLT network – created in the early 20th century, and an inspiration for later new towns – the freehold assets of the original development company, farmlands and all, have been incorporated into a charitable heritage foundation. Income is invested for long-term community benefit and generously supports a string of social ventures for the 33,000 inhabitants, complementing the welfare state.[13] Sadly, the subsequent new towns did not retain their assets. They were largely sold off by a Conservative government which also abolished the LSA in England, which once underpinned a network of smallholdings devoted to local food production, sometimes marketed nationally.

Communities first

The ethos of creating homes, and assets, for the community underpins other bold initiatives – often in the most unlikely places. Grayshott, in north-east Hampshire, could best be described as a prosperous commuter village of 2,500, with demand for housing outstripping supply. Teresa Jamieson knows all about that: the village intranet is buzzing with enquiries about the availability of local housing from restless Londoners keen to abandon the capital, she says – "it's almost continual". As chairman of one of the country's smallest housing associations, formed in 1935 to provide social homes for people on lower incomes, she's at the sharp end of trying to provide houses for those who could never afford to buy a home – sometimes key workers sustaining their better-off neighbours.

Grayshott and District Housing Association, run by volunteers, had 31 properties. But recently it was effectively reborn and embarked on its biggest scheme: taking out a loan of £1.6 million to build 14 flats.

Teresa, who formerly worked at a Citizens Advice centre, knows a thing or two about housing as a former chair of the local district council's housing committee. "The perception of the village is of affluence, people on high salaries," she acknowledges. "But there are still people here barely on the minimum wage." She mentions the self-employed, carers in a nursing homes sharing accommodation out of desperation, and even teachers in the local school who can't afford to live in the village.

But the flats are, deliberately, not classed as 'social housing', and for a reason. If they were, the association fears they could fall within the web of the government's continuing right-to-buy programme, which gives social tenants the ability to acquire their rented homes at discounts, whatever the misgivings of social landlords. Instead, in Grayshott, rents are calculated to be within 60–70% of a notional full market rent to avoid their being sold to sitting tenants. Teresa has seen the impact of the enforced sale of rented homes drastically reducing the stock of social housing: "dispiriting, to say the least", she laments.

Yet, opposition to new rural housing, social homes and all, is often strong, reinforcing what amounts to a rural–rural divide: the 'haves', with decent incomes, who've moved to the shires from cities, resenting any new development; and the 'have nots', born and raised in the countryside, with

8.1: Maintaining village schools is vital to underpin rural life: Cambo, Northumberland

8.2: Breath-taking beaches, like this in Kintyre, draw visitors – but remote places need shops and schools. The local village, Carradale, has the latter – but, until recently, it lacked the former

8.3: Carradale: the community raised enough funding for a store run by volunteers. It opened in June 2021

lower incomes, who don't earn enough to qualify for a mortgage – and can't find any social housing.

Take Cirencester, in Gloucestershire, and the surrounding Cotswold countryside, where locals – often incomers – have sometimes opposed new housing, no matter that the county had 17,000 people on the local council waiting list for social homes, thus presenting lower-income locals with two choices: leave, or rent privately. The views of a young couple from Cirencester, both with full-time jobs, are typical of many throughout rural England: "We were born here; we have a right to live here. They are making it so hard," says one. The other adds: "They basically want people with lots of money to come here, buy houses, and don't give a damn about any other people."[14]

Wooler and Grayshott may have similar populations, but their characteristics and social make-up are clearly different – not least the latter's role as a commuter village, 40 miles from London. But they share one problem: the absence of affordable housing, which both have been trying to address in their inventive ways. That's allied to the continued demand from city dwellers pushing up prices for rural housing, either for permanent homes or for holiday lets and second homes. Indeed, one of COVID-19's unexpected legacies may well prove to be the acceleration of urban flight and its consequent impact on rural housing markets. Need it be this way?

In truth, we shouldn't be where we are: three countries (Wales, Scotland, England) where rural house prices – "uniquely in the western world", says one expert – are substantially higher than those in towns and cities. The expert adds, with a hint of despair, that rural housing markets have become the vehicle for "an increasingly exclusive countryside". And an elderly countryside, too, with the average age in rural areas now 5.5 years higher than in urban areas.[15]

How did we end up like this? One problem is that England's rural policy has changed little since it was created after 1945, in an era of food shortages and rationing. As others have argued, in essence this policy was based on the containment of urban England and the protection of farmland and the countryside – 'socially regressive', in essence, by limiting rural development and keeping wages down, 'the product of an unholy alliance between farmers and landowners who politically controlled rural England and radical middle-class reformers who formulated post-war legislation'. This approach has continued to frame policy, according to the rural sociologist Mark Shucksmith, with further support from middle-class, home-owning incomers who have 'constituted an increasingly powerful vested interest in opposing development'.[16]

Whether the post-COVID-19 aftermath will herald unintended changes in the economy of rural Britain as commuting to work loses its attraction for many, is an open question. If a new economy, partly based on home working and rural hot-desking 'hubs', emerges more by accident than by design, the implications for town and village life could be profound, with pressure on schools, local housing markets and, inevitably, rural communications and road networks – potentially undermining a zero-carbon target of 2050.

But, of course, there's a possibility of another outcome – as it happens, built more positively around that 2050 net-zero target: the emergence of a green economy based on a nature-led recovery, and being actively pursued by NRW.

Landed – not stranded

Then there are the challenges, and the opportunities presented by remoteness, sometimes dramatic in scale and in ambition; none more so than on the western extremities of Scotland (and Britain). Here community enterprise operates, unashamedly, under the label of 'economic development': all this amid sandy beaches, high mountains, moorland covered with numerous freshwater lochans and a magnificent, rugged coastline intersected by fjord-style sea lochs.

This is the 64,000-acre North Harris estate, acquired for £4 million by a local trust in 2003, in one of Scotland's largest community buy-outs of private land.[17] Rich in wildlife, with Europe's largest population of golden eagles – and a growing colony of rarer sea eagles – the North Harris Trust, in the Western Isles, carefully balances nature recovery with job creation and – crucially – population retention in an area which has long suffered from migration. But here, again, something is stirring: people are staying, new affordable homes are emerging, businesses are starting and, remarkably, demand for new housing is outstripping supply, according to Michael Hunter, manager of the North Harris Trust, which has built four social homes already and is planning two more to complement two small, former council housing schemes owned by a non-profit housing association. As Michael underlines, housing is *the* economic issue on an estate with a population of 1,000, including the main settlement of Tarbert – with primary and high school, hotels, shops, ferry terminal to the – almost! – mainland (Uig, on Skye) busy harbour and a nearby island, Scalpay, connected by a bridge from Harris. Michael is realistic: "We need people, we have jobs from new businesses, but families cannot afford to buy – so we have to step in where possible." And they do. It helps, of course, that the trust owns what amounts to the costliest element in house building: land.

Of course, an economy originally built around crofting (smallholdings) and, hence, cooperation through sharing land – in this case 169 crofts and 29 areas of common grazing – is well suited to communal land ownership. It is helpful too that, unlike England, Scotland has abolished the Right to Buy for social housing stock – thus ensuring that affordable homes in the trust's area cannot be sold off. But it is helpful, most of all, because land reform legislation, in theory, makes it easier for community land buy-outs. And, assuredly, there'll be more of them.

But an asset on the scale of the vast North Harris estate still has to generate income, and employment, on top of the four full-time jobs in the trust itself, and others dependent on it. While green energy, from hydro and wind power, provides a modest electricity supply, the possibility of exporting this resource to the mainland clearly has potential, subject to grid capacity. With the estate given a top rating as a Site of Special Scientific Interest, allied to magnificent beaches and endless opportunities for hill walking and decent links to the mainland, the tourist possibilities are endless.

From her upland farm on the northern tip of Scotland, overlooking the Pentland Firth over to Orkney, Joyce Campbell also sees potential for 'repeopling' areas once sustaining a much higher population; carbon trading, tourism and (wind farm) energy all offer potential for growth. With a flock of around 830 breeding ewes, supplemented by 50 rams, a small herd of cattle and free-range hens, the family's 5,500 farm at Armadale, in Sutherland, is not heavily stocked. So far, locals have exploited energy to their advantage – gaining £350,000 from energy provider Scottish and Southern as compensation for allowing a local wind farm. The money was used to build a new community hall. Clearly, there's further potential – but there's also opposition from near neighbours in the form of Wildland Scotland, the land-holding enterprise of Danish billionaire Anders Holch Povlsen. Joyce well remembers a small demonstration when locals waved placards – "we are the local people: an endangered species" – outside an inquiry called to consider the case for more wind farms, opposed by a landowner.[18]

When a relatively new (Scottish) government agency, the Scottish Land Commission, notes that 'land is at the heart of Scotland's identity, economy and communities', you're reminded (in the words of the outgoing cabinet secretary for Environment, Land Reform and Energy, Roseanna Cunningham) that land reform in Scotland is a "process, not an event." It is ongoing. England take note![19]

Reading accounts of the remarkable efforts to encourage people onto the land from the 1920s through to the 1950s, in England, as well as in Scotland – with houses, small plots to cultivate and sometimes with agricultural training thrown in – you're reminded of a not-too-distant past when governments had the ambition, and drive, to create valuable rural jobs with purpose: feeding communities and, in the case of the former LSA, helping to feed the country with equally valuable crops produced by a new rural class.

Today, we face a similar challenge: feeding a nation over-dependent on food imports from the EU and beyond. During a virtual journey round

Britain in 2020–21, often retreading old, once-familiar ground, one theme has emerged above all others. It's articulated by organic pioneer Ian Tolhurst, in Oxfordshire, and countless other people and enterprises making a living from working the land: an absence of affordable housing to accommodate workers and fill emerging jobs in both food production and nature recovery. With a pandemic still casting an ominous economic, social (and health) shadow over Britain and the planet, here's one area with huge potential for employment growth in the great, healthy outdoors. It shouldn't be wasted.

Where, then, is that unbounded ambition, built on active government, which fired a Conservative–Liberal coalition after 1919, and then a Labour government after 1945 and, to some extent, after 1964? The countryside – the country – deserves better. And a housing policy built on reviving inclusive towns, villages and rural communities with practical solutions, rather than destructive dogmas, might be a start – along with a realistic appraisal of the measures needed to create turn-around towns and villages and, thus, help to rework the countryside.

9

Land Renewing: Reworking for All?

In nature there are neither rewards nor punishments; there are consequences.[1]

What keeps people working the land: growing crops, raising and trading livestock, enriching the countryside, nurturing nature, hoping to make a living from improving 'natural capital' – in plain English, embracing key resources such as soils, peatlands, water supplies, geology, wildlife and living organisms?

And how to assess the cost of renewing that most basic resource – *our* land – by reworking the countryside for the benefit of all to provide the food we need, enhance the landscape we love and address the climate emergency which threatens us all?

The average farmer's simple answer to the first question might well be "That's the only life I know". This invariably boils down to sentiment, comfort on home turf, being rooted to place, sometimes to language – certainly for farmers in the Fferm Ifan CIC. Their response to that question would be an emphatic "Because it's where we belong".[2]

The answer to the second question is more complex. It depends on a range of factors. Not least among these is the role of government in delivering an integrated land use strategy for England, learning – dare one suggest? – from the differing, and more coordinated, approaches of Scotland and Wales.

In this context, the argument from a former secretary of state for the Environment, John Gummer (Lord Deben), for a new department of land use to coordinate strategy across Whitehall is compelling. There's "no hope" of sensible land use in England, he maintains, when planning is "imprisoned" within the Ministry of Housing, Communities and Local Government (MHCLG), agriculture within the Defra, infrastructure within the Department of Business, Energy and Industrial Strategy, and transport within the Department for Transport (DfT).[3]

As it is, government departments in England are pulling in different directions – and never more so than when the ideology of an emerging, deregulatory planning regime promoted by the MHCLG collides with the worthy intentions of a more pragmatic Defra, which are further undermined by the carbon-heavy road-building strategy of the DfT.

Two months after Defra produced a detailed update of its Sustainable Farming Incentive (SFI), in March 2021, part of the ELMS regime to pay farmers to cut carbon emissions, MHCLG unveiled proposals which would, potentially, *increase* emissions. They wanted to make it easier for developers to build in rural England without planning consent under a new zonal system, side-lining local councils. When environmental campaigners – and, it must be added, leaders of Conservative councils – label this 'suburbanization' of the shires, encouraging 'rural sprawl', the consequences are clear: car use, in a countryside already with diminishing public transport (DfT's remit), will assuredly increase along with carbon emissions. New developments, after all, mean new and improved roads delivered by DfT, already committed to a big highway building programme.[4]

Defra's emerging ELMS regime is based on pricing nature as a basis for assessing future payments to farmers, partly through the SFI. Defra has already calculated that the SFI will 'help to protect and enhance' an estimated nine million tonnes of carbon already 'stored' in England's hedgerows. They say this, alongside other environmental proposals to reduce greenhouse gas emissions, is 'roughly the equivalent' of taking 40,000 cars off the road between 2023 and 2027 and up to 130,000 between 2028 and 2032.[5]

You get the point: at best, these are colliding proposals. This confusion – departments operating at cross-purposes, with little coordination – is replicated across Whitehall and (English) government. Lord Deben, chairs the CCC, a government body created through the Climate Change Act 2008. Areas from planning and the environment to agriculture and infrastructure make a "cohesive whole", he insists. Policy coherence is therefore vital "as we face up to the huge changes we will have to demand from our farmers to deal with flooding and climate change as well as the depletion of the fertility of our soils".[6]

As it is, we are a long way from the intentions of a White Paper produced 77 years ago – remarkably, the last time land use was considered holistically by a UK government (long before Scottish and Welsh devolution). Titled *The Control of Land Use*, it talked inspiringly about the need to gain the 'best use of land in the national interest'.[7] No surprise, then, that after 1945 a raft of legislation followed: from the Town and Country Planning Act 1947 (now being dismantled in England), to the National Parks and Access to the Countryside Act two years later, which delivered ten National Parks and legally underpinned almost 140,000 miles of footpaths (still enduring).

With the benefit of hindsight, it's perhaps easier – or, over charitable, depending on your viewpoint – to explain policy underlining a third piece of legislation which unleashed the intensification of farming – the

Agriculture Act 1947 – as an aberration. Consequently, in today's context, the words of the then Minister of Agriculture, Tom Williams – "to promote a healthy and efficient agriculture" – might ring a little hollow. But this was the aftermath of war; the government was desperate to achieve a positive balance of payments, reduce imports and make Britain self-sufficient in food. This it almost achieved. Whether near-self-sufficiency could have happened without unleashing chemicals, fertilizers and insecticides on the countryside – ripping down hedgerows to create big field systems in the process – is an open question. That debate will continue.

But while self-governing Scotland and Wales tread a different path today – broadly sticking to the principles of the 1947 planning legislation, for instance – England is moving in a distinctly different, ideologically driven direction.[8] To be fair, the ELMS regime, still to be fully worked out, sits alongside an Environment Bill providing legal frameworks for water quality, nature conservation and air pollution. Potentially, it points to a progressive way forward, no matter how limited and belated; it's five years, after all, since the 2016 Brexit referendum signalled the end of the EU's much-derided CAP regime and, hence, a new direction for farming, food and land use.

As it is, we are facing the most radical changes in land use for over 70 years with no clear direction of travel in England; certainly, nothing, for instance, to match the scale and ambition of that 1944 White Paper. As Baroness (Barbara) Young, former chief executive of the EA argues, changes in farming support – namely, the ELMS regime – must ensure that the urgent task of reversing declines in once-common wildlife is built into an "integrated approach to land use and practice for agriculture". The one silver lining of Brexit, she says, is the opportunity to build an integrated land-use strategy. We are still waiting. She suggests learning from Scotland and Wales.[9]

And so to food

The imperative, surely, should be to keep us adequately fed through nature-friendly farming by increasing domestic crop production to underpin food sovereignty; making Britain more resilient to withstand the uncertainties of imports, which provide 40% of our food overall.[10] All that involves joined-up government to persuade farmers, and landowners, to reverse long-standing practices in agriculture and in land management – addressing the climate emergency by reducing greenhouse gas emissions – and accepting that pursuits such as shooting grouse, a moorland bird curiously labelled 'game', should be curtailed in the interests of restoring peatland to a carbon-rich state.

In England, a centrepiece of the new Agriculture Act, the emerging ELMS regime embracing sustainable farming, is certainly a start. But it depends on multi-billion-pound funding annually from the Treasury to succeed – largely replacing farming subsidies, which make up more than half of farm incomes.[11] The current formula, based on giving the highest subsidy to those with the most land – now being phased out – turns economic logic on its head. It had to go.

Life outside the EU's CAP regime, then, presents a heaven-sent opportunity to renew and rework the countryside for the benefit of everyone – people seeking solace, inspiration, rest, recreation, nourishment, spiritual renewal, as well as those working the land – with the most profound changes in the management of our wide open spaces in generations.

Agriculture, an industry surely with potential for expansion at both community level and in new, technologically driven areas – indoor vertical and hydroponic farming, for instance – will always need some government support if domestic food production is to rise above the current low level of self-sufficiency. A UK blessed with so many acres devoted to farming – 72% of its land, one of the highest levels in Europe – and with the added potential of multiple harvests under glass should surely be able to feed itself.

While a long-overdue emphasis on environmental land management is welcome – not least, to reverse the impact of intensive farming – it must not come at the expense of increasing domestic food production. In all this, it's fair to ask if England's emerging ELMS regime can ever reach its worthy objectives, for two reasons: firstly, the money is unlikely to be available to replicate outgoing subsidies, currently over £3 billion annually; and secondly, there's no agreed national or international formula on how to value nature and how to compensate farmers and landowners for enhancing it. Not yet, anyway. But many fine minds are working on it globally.[12]

Business is stirring too under a new label – agritech – and a fast-growing investment trend for start-ups aiming to apply new scientific techniques, and AI, to improve productivity by farming sustainably through natural processes. Largely under the radar, these start-ups have been targeted by 'special purpose acquisition companies' – 'so-called blank cheque companies that first list on the stock market and only then find [another] company to purchase'.[13] If the government, on the back foot, remains passive on this front, City investors certainly are not. They smell money. One income fund, for instance, in highlighting the "considerable" tax benefits from investing in forestry – it is already targeting 24,700 acres of new woodland – further underlining the added potential from selling carbon credits.[14]

In self-governed Scotland, a distinctly different agenda is emerging. Driven by a campaigning charity, Community Land Scotland, pressure is growing for the new Parliament (elected in early May 2021) to approve legislation giving ministers powers to stop the sale of large estates, and to break up existing land monopolies if they are acting against the interest of local communities. A government agency, the Scottish Land Commission, appears sympathetic but it's not clear whether the SNP government favours such radical action. Nevertheless, in moves that seem unimaginable in England, Scottish ministers also want to double community land ownership to well over one million acres.

Yet England – here's room for cautions optimism – is providing imaginable steps on other fronts. Some farmers and landowners are already taking measures to improve soil quality, reduce or eliminate harmful chemical fertilizers and pesticides and rewet peatland and moors drained for sheep grazing and grouse shooting; in short, they are significantly changing land-use practices.

And then, of course, there are those 'new revolutionaries' and other long-standing advocates for nature-friendly farming. A relatively new movement is, literally, growing – and the sum of its parts adds up to an emerging network starting to drive change, providing jobs, foreseeing further employment growth, creating local supply chains and trading directly with consumers; field to fork, so to speak. But there's one missing ingredient: access to land. Time and again, this new wave of smaller-scale farmers stress that they could achieve so much more if land were made available. As organic pioneer Ian Tolhurst says, there's "definitely a growing movement [of local producers] … trying to look after neighbourhoods, with a high degree of resilience". With a hint of despair, he adds: "They don't have access, to land, training … learning as they go along … but they do have a direct link to the consumer." Sir Julian Rose, whose family owns the Hardwick estate in Oxfordshire, on which Ian farms, adds: "We need other landowners to take the same kind of approach as I'm taking to gain equilibrium … [be] a little bit more flexible and humanitarian … less protective … freeing up small acreages for people."

Clearly, smaller-scale producers are in no position to rival the Big Four supermarket chains, whose business model is built on vulnerable, just-in-time delivery systems: from grower or producer and importer to distribution depot and, hence, to stores, with endless food miles in between. "But we're much more nimble," insists Tolhurst.

Growing with confidence

Take one inspiring example. A meeting of adventurous farmers in January 2018, at the Oxford Real Farming Conference, led to the emergence of a new movement: the Nature Friendly Farming Network (NFFN). From that small beginning, it now has 2,000 farming members, 7,000 individual signed-up supporters, and represents 125 organizations spread throughout the UK, transcending the borders with Wales and Scotland. That, says Martin Lines, its chair, should tell us something.[15]

Martin has transformed his family's Cambridgeshire arable farm into a productive holding by working with nature, ending all ploughing – which releases carbon – and seeding through drilling the soil instead, renewing and planting hedgerows, creating wild meadows and successfully encouraging bird life with generous quantities of specialist seed to attract snipe, woodcock, reed and corn buntings, yellowhammers – to name a few – and, of course, skylarks. "It's joyful," he enthuses.

Martin Lines, a third-generation farmer, decided over ten years ago to transform the family's 420-acre Papley Grove Farm, supplemented by 500 rented acres, into what is now a showpiece for the NFFN. Out went 'traditional' intensive practices, chemically driven, no matter the consequences for wildlife, plants and wild flowers; in came regenerative farming, few if any artificial fertilizers, underpinned by a commitment to biodiversity and working with nature to recover some of the estimated 600 farmland species which have declined dramatically since the early 1970s.[16] The aim is to renew his most important asset – the soil – and help the family business to become more profitable and maintain productivity. And so it has transpired.

Alongside friends from the NFFN, Martin is similarly passionate about recreating local food networks, with direct farm-to-table supply chains: "I strongly believe they'll come back," he maintains. And those of us of a certain age do remember a time when council markets (in my case, Carlisle) and other outlets were effectively local food 'hubs', groaning with veg and salad crops from nearby market gardens (remember them?).

This raises a leading question: if a 'remarkable coalition of forces, interests and organisations, a broad and new consensus' emerged in the 1930s to argue that land should be better used to ensure people were adequately fed – this at a time when barely a third of our food was home grown – why not a similar movement today, when we're lucky to produce 60% of the crops we need?[17] Hopefully, as Martin says, something *is* stirring.

New beginnings?

Yet, a vital element is missing: an absence of urgency by a UK government, bereft of any food policy, still less an overarching strategy to address land use and, hence, the climate emergency. Consider the array of organizations below the government department in England – Defra – ostensibly overseeing strategy and delivery: there's the EA, charged with river quality, flood and coastal defences; Natural England, overseeing – as the name implies – nature and its recovery; the Rural Payments Agency, which distributes farming subsidies (and, presumably, their replacement, the ELMS scheme); and soon there will be the Office of Environmental Protection, meant to oversee functions – pollution, air quality, for instance – previously undertaken by the EU. Then there's the CCC, chaired by Lord Deben, unafraid to speak truth to power. What's needed is a more powerful body with, maybe, some

9.1: Quietly flows the river: tranquillity on the North Tyne, Wark, Northumberland

9.2: Late spring in hill country: the North Pennines area of outstanding natural beauty

9.3: Public rights of way: a legacy of the National Parks and Access to the Countryside Act 1949

delivery functions, charged with advising a UK government which, so far, has turned its back on critical friends. After all, a Countryside Agency in England, created in 1999 and partly absorbed into a Commission for Rural Communities (abolished in 2013), operated successfully as a delivery and then as a research and policy body – ensuring that the countryside had a relatively high priority. At the very least, an agency – which could, absorb, perhaps Natural England, elements of the EA, and some functions from Defra – might make a stab at coordinating a countryside and land-use strategy across government. As it is, others are taking the lead.

If that journey from 'intensive' to nature-friendly farming proved a challenge for Martin Lines and the NFFN movement, think of the impact the new environmental agenda, implicit in the ELMS regime, will have on the country's largest (non-governmental) landowner, the National Trust, whose 610,000 acres cover parts of England, Wales and Northern Ireland; 60% of them are let as whole farms. It has a commitment to reach a carbon net-zero target by 2030 across what Patrick Begg, the Trust's Director of Natural Resources, calls "our whole value chain" as a matter of urgency – and the impact across its varied holdings, from 780 miles of coast, to mountain, moor, pasture and 500 historic houses could be profound. Begg is adamant that the Trust, as a conserver of public assets – and constitutionally embedded in a specific Act of Parliament – has to show a national leadership role, including the carbon footprint on its 2,000 tenanted farms. This will require "big decisions and trade-offs"; in other words, winners and losers – not least, perhaps, sheep-stocking levels on the hills and the uplands – as part of broader picture in which land renewal and natural restoration are a primary focus. "We've given ourselves a rightly big task," Begg acknowledges.[18] Others are sceptical, fearing the Trust – England's biggest farming landlord – will inevitably trim its livestock farming operations and reap a potential new income stream from the emerging ELMS regime. How many tenant farmers will survive is an open question.

Subsidy woes?

As the Farming Minister in England, Victoria Prentis, has tacitly acknowledged, putting a price tag on improvements in nature to deliver the new ELMS subsidy for farmers is proving "not easy".[19] Indeed, so complex is the delivery of the first 'component' in a three-tier system – the SFI – that the process, says the minister, will take some time. Such is the delay, and obfuscation, that it's hard to escape the conclusion that the complexity of a pricing regime might prove a challenge too far; that ministers could be forced to fall back on a version of the outgoing CAP regime, under a 'sustainable' label, calculated according to either the acreage farmed or 'income foregone'. As one respondent to a ministerial interview on the

BBC's *Farming Today* programme commented: "My heart sank. Brexit was meant to bring clarity."[20] On some estimates from leading agricultural consultants, farming subsidies will fall by 50–60% under the ELMS scheme – compared with more generous grants from the outgoing CAP regime – even for those delivering the maximum level of nature-friendly 'services', through enhancing flora and fauna. For others, the picture looks bleak.

To be fair, Martin Lines thinks a scheme which compensates farmers for improving soil and water quality, planting trees and restoring peatland, for instance, can be made to work – but it seems that Defra may well have to take on trust 'self-declared information' on nature-recovery projects, right down to farmers submitting smartphone photographs as proof of nature-friendly work undertaken.

All this might seem an academic debate, far removed from reality on the ground. Emphatically, it is not. This is, literally, about the ground beneath our feet: the degradation of the soil, the source of all life. Up to now, this has appeared of secondary importance to the climate crisis. We needed 300 scientists to remind us in a UN Food and Agriculture Organisation (FAO) report late in 2020 that the worsening state of soils is at least 'as important as the climate crisis and destruction of the natural world above ground'. The words of a lead author of the report, Professor Richard Bardgett, of the University of Manchester, should be ringing in our ears: "There is a vast biodiversity living in the soil out of sight and generally out of mind. But few things matter more to humans ... we rely on the soil to produce food."[21]

Subsidy – or support?

Up to now, farming has been defined by one characteristic: unprofitability and, in some cases it has been argued, near-irrelevance. After all, just under half of England's farms produce 2% of total agricultural output, while 8% of them account for over half of it. Of course, larger farms generally deliver a surplus, and the very largest, such as Sir James Dyson's substantial operation in Lincolnshire, say their business model is built on operating without a subsidy – although Dyson's operation has received £8.7 million in the EU's CAP subsidy since the EU referendum in 2016.[22]

The hefty sums paid out to the largest landowners expose deeper issues of inequity when smallholders with land of less than 12 acres receive nothing – although, collectively, they add up to an alternative food movement of community producers, sometimes providing jobs and creating and sustaining local supply chains.

If the UK government remains passive, what's encouraging, as we saw earlier, is the ability of others to take matters into their own hands. For some, kinship ties to the land are so strong that those 11 farmers in the

Fferm Ifan CIC in Snowdonia have already, through self-help and support from the Welsh Government, made a good stab at renewing their land so that ministers in Cardiff can put a price on their labours.[23]

In the lowlands as well as the uplands, farmers already either improving landscapes or farming in a nature-friendly way – sometimes called 'regenerative farming' – are ahead of the curve. Others are not. They want out. That much was clear in a revealing survey undertaken by the TFA among 360 of its members to gauge whether they want to stay in the industry as agricultural subsidies, a legacy of our EU membership, are phased out.[24]

Three-quarters of respondents said they were 'seriously interested' in quitting the land and taking a 'golden goodbye' promised by Defra: officially, a Lump Sum Exit Scheme based on an aggregation of annual subsidies from 2021 to 2028, calculated on the amount of land farmed. Although consultations have been promised, some wise heads are suggesting that the government might find it politically challenging to approve all but a modest scheme, in the light of heavy job losses elsewhere in the economy.

Still, many farmers are clinging on – just. They are rooted to the land, whatever the losses they might incur; but many, of course, are tenants, and will still have to pay rent to a landlord, whether the National Trust, the Prince of Wales's Duchy of Cornwall estates, offshore-registered trusts created as tax-efficient vehicles for aristocratic families or wealthy individuals seeking similarly legal tax 'breaks'. 'Many farmers put the farming lifestyle as being more important to them than maximising profits,' Defra has noted.[25] Which is fine if you own a farm; rather less so if you're a tenant; yet many, particularly in the hills and uplands, remain as 'hefted' to their (rented) land as their sheep grazing the hills.

As it is, 48% of farms in England are either wholly or partly rented, and 52% are owner-occupied, although many farmers fall between these two categories, owning some land and renting more. But, as Defra has noted 'there is evidence that direct payments [from EU subsidies] have inflated farm rents' – thus giving landlords easier (taxpayer-funded) pickings, while some tenants struggle to pay, sometimes living almost from hand to mouth.[26] Consequently, with direct subsidies now being phased out, it's fair to ask – and Defra has raised the question – if rents for hard-pressed tenants will now be reduced. For owner-occupier farmers, it's a different story; according to Defra they are 'asset rich', with an average net worth of almost £2 million each.[27]

What's clear, however, is that the EU's generous subsidy has 'effectively hidden' hill and upland farmers from financial reality, cushioned by the CAP. Broadening this out to England, the National Audit Office found in 2019 that 42% of the country's farmers made a loss over the previous two years.[28]

But, at the same time, many of the landscapes in these areas, embracing what agronomists call 'ecosystem services' – what you and I call water, soils and habitats for wildlife – are in 'poor condition'. What's more, stocking levels for sheep have been kept unnaturally high, leading to so much over-grazing that additional feedstuffs are needed because there's insufficient grass on which to feed – compounding the poor condition of the land. With detailed analysis, a revealing report on the plight of these hill farmers concludes that there's a case for survival if they reduce sheep numbers and exploit potential payments from the emerging ELMS regime in England.[29]

But, undoubtedly, some will not survive. Whether there's a significant flight of farmers from the hills, particularly among tenants – who dominate many upland areas, sometimes with tenancies across generations – remains to be seen. But the vision of an upland transformation, from hill farming to 'green prairies', should perhaps be taken seriously: namely, the prospect of current owners, or investors repurposing or acquiring large tracts of land vacated by farmers and qualifying for substantial payments from the ELMS, selling on carbon credits to, say, food retailers and transport companies to reduce their carbon footprint.

Britain, then, is entering a brave new world across its varied countryside: a transformation, potentially a quiet revolution signalling far-reaching changes in land use and farming support.[30] In the preceding chapters I've argued that striking a balance between efficient, nature-friendly farming on the one hand and managing the land and its fragile habitats and ecosystems on the other need not always involve hard choices. The two, after all, are one of the same. They coexist.

Reality check

But here's the reality. We should no longer expect others to feed us. We have to act. There is hope; the preceding chapters have underlined the enthusiasm, the inventiveness, the potential for growth among growers and producers expanding against all the odds – and, it must be said, the pragmatism of some of the largest agricultural enterprises transforming their farming practices. We need, in short, a balance: supporting small-to-medium enterprises, sometimes local cooperatives, doesn't necessarily mean rejecting either the mega-farms or the emerging large-scale producers growing crops in multiple harvests under glass.

One ingredient is missing: joined-up government. On a shared island facing so many challenges – addressing the climate emergency; low self-sufficiency in food; the threat to our best farmland in the Fens, and to coastal communities; the urgency for peatland and upland renewal – the case for a cohesive strategy in land use across (English) government needs addressing with urgency. While a new Department for Land might be the ideal, resurrecting a specialist rural or countryside agency in England, perhaps by absorbing other bodies, like Natural England – in short, an advisory body, with added delivery powers over land and natural resources – would be a start.

But an extra dimension is needed; in territorial terms, we can learn so much from each other. England might be the largest country in a UK of three nations, and a share of a fourth – but in framing progressive policy, joining up strategy, it is, frankly, an outlier. If we share the same island, eat the same food from the same fields (or poly-tunnels and glasshouses), share the same distribution networks, belong often to the same trade or representative organizations, then surely we need a forum in which to talk and, where appropriate, to jointly act in an alliance some call Britain.

Already, from the inspiring growers literally at the grassroots in countless communities, to the renewed fields in smaller farming enterprises and larger operations and, hopefully, to the uplands and the degraded moorlands, something is moving. While it's premature to call it the makings of a new rural economy in which farming is a part, a new wave of growers, producers, 'renewers' – activists, yes – provide at least a little hope for a brighter future in rural Britain, in spite of the seeming indifference among much of the English political class.

We could start with modest reforms: a new charter for land, embedded legally in titles, for instance, setting out rights and responsibilities for farmers and for landowners. And we should address one element which is key to beginning full-scale reform – ending the inheritance and capital gains tax breaks which make land trading so attractive for the few at the expense of the many. As the former first secretary of the UK's HM Revenue and Customs said in 2021, the "massive relief" in these taxes for agriculture and business "do not have any credible economic argument" – although abolishing inheritance tax relief, particularly, would "get squeals" from those affected.[31]

Far from being radical reform, these measures would represent modest, practical steps for a market economy to at least begin addressing a global and a national crisis – the climate challenge, allied to food insecurity – driven by an activist state overseeing regulated markets underpinned by competition rather than by inheritance, unearned riches and, hence, by undue landowner dominance.

Our land, while facing unprecedented pressures, also presents boundless opportunities for renewal. As our most basic, life-giving resource, it should be nurtured, treasured, truly valued for its foundational worth, rather than for purely monetary gain. It is, after all, our most valuable asset. Reworking it should provide the means to deliver so much more – in food, employment, leisure and, ultimately, in enrichment and enjoyment for all. In short, land *renewed*.

Notes

Introduction

[1] Hetherington, P. (2020) www.theneweuropean.co.uk/brexit-news/shopping-local-makes-coronavirus-crisis-comeback-73896.

Chapter 1

[1] Proverbs 2:20–22

[2] Ingram cafe and museum, www.searoundbritain.com

[3] Defra/Government Statistical Service (2019) *The Future Farming and Environment Evidence Compendium*, https://defra.gov.uk

[4] Defra/Government Statistical Service (2019) *The Future Farming and Environment Evidence Compendium*, https://defra.gov.uk

[5] Defra/Government Statistical Service (2019) *The Future Farming and Environment Evidence Compendium*, https://defra.gov.uk

[6] Lang, T. (2021) Interview with author (May); Lang, T. (2020) *Feeding Britain: our food problems and how to fix them*, London: Pelican

[7] Lang, T. (2021) Interview with author (May)

[8] Ingram cafe and museum, www.searoundbritain.com

[9] HM Government (1944) *The Control of Land Use*, White Paper (Cmd 6537)

[10] Defra/Government Statistical Service (2019) *The Future Farming and Environment Evidence Compendium*, https://defra.gov.uk

[11] Young, B. (2021) What Whitehall might learn from Scotland and Wales, *Town and Country Planning* (March/April), www.tcpa.org.uk

[12] CPRE: the countryside charity (2017); various contributors. *Landlines: why we need a strategic approach to land.* https://www.cpre.org.uk

[13] Lang, T. (2021) Interview with author (May)

[14] Hubbard, C. (2019) *Brexit: How might UK agriculture thrive or survive?* ESRC, Newcastle University Centre for Rural Economy, www.research.ncl.ac.uk

[15] Hetherington, P. (2015) *Whose Land is Our Land: the use and abuse of Britain's forgotten acres*, Bristol: Policy Press

[16] Zephaniah, B. (2020) *Space for Nature: The UK in one hundred seconds*, www.friendsoftheearth.uk

[17] Cahill, K. (2002) *Who Owns Britain*, Edinburgh: Canongate

[18] Strutt and Parker (2021) Scottish Estate Market Review/England Estates and Farmland Market Review (Spring), https://rural.struttandparker.com/article/record-sums-invested-in-scottish-estates-during-2020/; https://rural.struttandparker.com/article/green-investment-in-farmland-rises/

[19] Strutt and Parker (2021) Scottish Estate Market Review/England Estates and Farmland Market Review (Spring), https://rural.struttandparker.com/article/record-sums-invested-

in-scottish-estates-during-2020/ https://rural.struttandparker.com/article/green-investment-in-farmland-rises/

20 Lowe, P., Marsden, T., Munton, R. and Flynn, A. (1993) *Constructing the Countryside*, London: UCL Press Ltd

21 Ellis, H. and Henderson, K. (2014) *Rebuilding Britain: planning for a better future*, Bristol: Policy Press

22 Hetherington, P. (2015) *Whose Land is Our Land: the use and abuse of Britain's forgotten acres*, Bristol: Policy Press

23 Lowe, P. et al (1986) *Countryside Conflicts: the politics of farming, forestry and conservation*, Aldershot: Gower Publishing

24 Lang, T. (2021) Interview with author (May)

25 Lowe, P. et al (1986) *Countryside Conflicts: the politics of farming, forestry and conservation*, Aldershot: Gower Publishing

26 Kynaston, D. (2007) *Austerity Britain 1945–51*, London: Bloomsbury

27 Newby, H. (1988) *The Countryside in Question*, London: Hutchinson

28 Lewis, D. (2020) The UK's food strategy cannot be left to the market, *Financial Times*, 16 July, www.ft.com/content/acfd1251-24e9-478b-b8db-c6cc70d64b39

29 Carrington, D. (2019) Convert half of farmland to nature, urges top scientist, *The Guardian*, 31 December

30 Climate Change Committee (2020) *Agriculture and Land Use, Land Use Change and Forestry*, Sixth Carbon Budget

31 Rebanks, J. (2020) *English Pastoral – An Inheritance*, London: Penguin Random House

32 Newby, H. (1980) *Green and Pleasant Land? Social change in rural England*, Harmondsworth: Penguin

33 Shucksmith, M. (2018) *Rural Policy after Brexit*, Academy of Social Sciences, Routledge

34 Shucksmith, M. (2018) *Rural Policy after Brexit*, Academy of Social Sciences, Routledge

35 Wilson, S. (2005) *Reflections: The Breamish Valley and Ingram*, Newcastle-upon-Tyne: Northern Heritage

36 Hetherington, P. (2020) Farmageddon – how the countryside is being shafted by Brexit leaders, *The New European*, 5 March, www.theneweuropean.co.uk

37 Lang, T. (2021) Interview with author (May); Lang, T. (2020) *Feeding Britain: our food problems and how to fix them*, London: Pelican

38 Hetherington, P. (2015) *Whose Land is Our Land: the use and abuse of Britain's forgotten acres*, Bristol: Policy Press

39 Lang, T. (2021) Interview with author (May); Lang, T. (2020) *Feeding Britain: our food problems and how to fix them*, London: Pelican

40 Hetherington, P. (2015) *Whose Land is Our Land: the use and abuse of Britain's forgotten acres*, Bristol: Policy Press

41 Hetherington, P. (2015) *Whose Land is Our Land: the use and abuse of Britain's forgotten acres*, Bristol: Policy Press

42 Wilson, S. (2005) *Reflections: The Breamish Valley and Ingram*, Newcastle-upon-Tyne: Northern Heritage

43 Carrington, D. (2019) Convert half of farmland to nature, urges top scientist, *The Guardian*, 31 December

44 Hetherington, P. (2015) *Whose Land is Our Land: the use and abuse of Britain's forgotten acres*, Bristol: Policy Press

Chapter 2

1 Lloyd George, D. (1909) Speech in Limehouse, East London, 30 July

2 Lloyd George, D. (1909) Speech in Limehouse, East London, 30 July

3   Hetherington, P. (2019) Land, Property and Reform. In *The Routledge Companion to Rural Planning*, Abingdon: Routledge

4   Briggs, S. (nd) Agricology, www.agricology.co.uk/field/farmer-profiles/stephen-briggs

5   Hetherington, P. (2019) *Land of Plenty: What agricultural land could do for councils*, https://publicfinance.co.uk

6   Cambridgeshire County Council (2021) Commercial and Investment Committee, www.cambridgeshire.gov.uk

7   Cambridgeshire County Council (2021) Commercial and Investment Committee, www.cambridgeshire.gov.uk

8   www.norfolk.gov.uk/business/business-development-opportunities/county-farms

9   Hetherington, P. (2019) *Land of Plenty: What agricultural land could do for councils*, https://publicfinance.co.uk

10 Dearlove, P. (2007) *Go Home You Miners! Fen Drayton and the LSA*, Pamela Dearlove

11 Hall, P. and Ward, C. (2014) *Sociable Cities: The 21st-Century reinvention of the garden city* (2nd edn), Abingdon: Routledge

12 Dearlove, P. (2007) *Go Home You Miners! Fen Drayton and the LSA*, Pamela Dearlove

13 Dearlove, P. (2007) *Go Home You Miners! Fen Drayton and the LSA*, Pamela Dearlove

14 Dearlove, P. (2007) *Go Home You Miners! Fen Drayton and the LSA*, Pamela Dearlove

15 Defra/Government Statistical Service (2019) *The Future Farming and Environment Evidence Compendium*, https://defra.gov.uk

16 Dearlove, P. (2007) *Go Home You Miners! Fen Drayton and the LSA*, Pamela Dearlove

17 Lang, T. (2020) *Feeding Britain: Our food problems and how to fix them*, London: Pelican

18 Hall, P. and Ward, C. (2014) *Sociable Cities: The 21st-Century reinvention of the garden city* (2nd edn), Abingdon: Routledge

Chapter 3

1   Schumacher, E.F. (1973) *Small Is Beautiful: A study of economics as if people mattered*, Chapter 7 The proper use of land, HarperCollins

2   Rose, J. (2020) Interviewed by author

3   Tolhurst Organic Partnership CIC, https://tolhurstorganic.co.uk

4   *The Guardian* (2020) City dwellers idealise Britain's countryside, www.theguardian.com/society/2020/oct/13/city-dwellers-idealise-britains-countryside-but-theres-no-escaping-rural-poverty, 13 October

5   Shelter (2004) *Priced Out: The rising cost of rural homes*, www.shelter.co.uk

6   Schumacher, E.F. (1973) *Small is Beautiful: A study of economics as if people mattered*, Chapter 7 The proper use of land, HarperCollins

7   www.landworkersalliance.org.uk

8   Lang, T. (2020) *Feeding Britain: Our food problems and how to fix them*, London: Pelican

9   www.landworkersalliance.org.uk

10 www.landworkersalliance.org.uk

11 Lang, T. (2020) *Feeding Britain: Our food problems and how to fix them*, London: Pelican

12 Lang, T. (2020) *Feeding Britain: Our food problems and how to fix them*, London: Pelican

13 Lang, T. (2020) *Feeding Britain: Our food problems and how to fix them*, London: Pelican

14 www.landworkersalliance.org.uk

15 www.naturalresourceswales.gov.uk

16 Prince of Wales (2020) *The Prince of Wales takes part in BBC's Rethink Series*, www.princeofwales.gov.uk

17 Evans, J. (2020) Britain's farmers braced for post-Brexit subsidy gap, www.ft.com/content/81009ae6-b825-49e4-8b95-c68cdf2a69b6

18 Shucksmith, M. (2018) *Rural Policy after Brexit*, Academy of Social Sciences, Routledge, https://doi.org/10.1080/21582041.2018.1558279

Chapter 4

1   1 Ecclesiastes 11:4
2   Dyson, J. (2020) Technology and Farming: Efficient high-tech agriculture (September), www.beeswaxdyson.com
3   NFU (2015) *Backing British Farming in a Volatile World*, www.nfuonline.com/635-15tl-the-report-digital-low-res/
4   Grosvenor Farms. Dairy Farming, www.grosvenorfarms.co.uk/is-it-possible-to-produce-high-quality-nutritious-milk-and-cereal-grains-while-managing-the-land-sustainably-to-restore-the-environment-and-support-biodiversity
5   Juniper, T. (2021) Interview with author (March)
6   Low carbon farming, www.lowcarbonfarming.co.uk/the-crown-point-estate/
7   Dyson, J. (2020) Technology and Farming: Efficient high-tech agriculture (September), www.beeswaxdyson.com
8   www.theneweuropean.co.uk/brexit-news/dyson-biggest-beneficiary-of-eu-subsidy-despite-brexit-96192
9   Defra/Government Statistical Service (2019) *The Future Farming and Environment Evidence Compendium*, https://defra.gov.uk
10 Stanley, J. (2020) Interview with author
11 Dunn, G. interview with author; www.tfa.org.uk; Defra/Government Statistical Service (2019) *The Future Farming and Environment Evidence Compendium*, https://defra.gov.uk
12 Dunn, G. interview with author; www.tfa.org.uk; Defra/Government Statistical Service (2019) *The Future Farming and Environment Evidence Compendium*, https://defra.gov.uk
13 Fewings, A. (2020) Interview with author
14 Crown Estate, annual report, www.lowcarbonfarming.co.uk/the-crown-point-estate/
15 Hintze, M., www.michael.hintze.com
16 Hetherington, P. (2015) *Whose Land is Our Land: The use and abuse of Britain's forgotten acres*, Bristol: Policy Press
17 Humphries, W. (2017) Duke of Westminster's £8bn fortune escapes death duties. *The Times*, 13 October, www.thetimes.co.uk
18 Hetherington, P. (2015) *Whose Land is Our Land: The use and abuse of Britain's forgotten acres*, Bristol: Policy Press
19 Tax Justice Network, www.taxjustice.net
20 Hamilton, S. (2020) s.hamilton@stephenson.co.uk
21 Hetherington, P. (2015) *Whose Land is Our Land: The use and abuse of Britain's forgotten acres*, Bristol: Policy Press
22 Williamson, R. (2020) Interview with author; www.beeswaxdyson.com
23 Dyson, J. (2020) Technology and Farming: Efficient high-tech agriculture (September), www.beeswaxdyson.com
24 Lang, T. (2020) *Feeding Britain: Our food problems and how to fix them*, London: Pelican
25 Briggs, S. (2010) Interview with author
26 Rebanks, J. (2020) *English Pastoral: An inheritance*, London: Allen Lane
27 Fiennes, J. (2020) Interview with author; The Holkham Farming Company, www.holkham.co.uk
28 Prentis, V. (2020) Interviewed on *Farming Today*, BBC Radio Four, 11 November
29 Dyson, J. (2020) Technology and Farming: Efficient high-tech agriculture (September), www.beeswaxdyson.com
30 Price, F. (2020) Interviewed by author

[31] Rebanks, J. (2020) *English Pastoral: An inheritance*, London: Allen Lane
[32] Chester Master, H. (2020) Interviewed by author
[33] Batters, M. (2020) Interviewed on *On Your Farm*, BBC Radio Four, 22 August

Chapter 5

[1] Wordsworth, W. (1815) To Sleep, *Poems*, Vol 2
[2] Dunning, J. (2020) Interviewed by author
[3] Dunning, J. (2020) Interviewed by author
[4] Dunning, J. (2020) *Westmorland Yeoman*, Kendal: Hayloft Publishing Ltd
[5] *Hansard* (1969) vol 786, North Pennines Rural Development Board, 11 July
[6] Lloyd, J. (2020) Interviewed by author
[7] Monbiot, G. (2020) Interviewed by author
[8] Rebanks, J. (2020) *English Pastoral: An inheritance*, London: Allen Lane
[9] Stocker, P. (2020) Interviewed by author
[10] Defra/Government Statistical Service (2019) *The Future Farming and Environment Evidence Compendium*, https://defra.gov.uk
[11] Defra/Government Statistical Service (2019) *The Future Farming and Environment Evidence Compendium*, https://defra.gov.uk
[12] Taylforth, E. (2020) Interviewed by author
[13] Churchouse, M. (2020) Interviewed by author
[14] Hunter, J. (2020) Interview; www.fernhill-farm.co.uk
[15] Prince of Wales (2020) Campaign for Wool 10th anniversary, www.campaignforwool.org
[16] *Farmers Guardian* (2020) Call for wool use in home insulation, 31 July
[17] British Wool (2020) www.britishwool.org.uk
[18] Harris Tweed (2020) Interview, www.harristweedhebrides.com
[19] Natural Fibre Co (2020) Interview, www.thenaturalfibre.co.uk
[20] Milton, R. (2020) Interviewed by author

Chapter 6

[1] Clare, J. (1824) *The Fens*
[2] Defoe, D. (1724–27) *A tour through the whole island of Great Britain divided into circuits and journeys*, Letter 1, Part 3, Norfolk and Cambridgeshire
[3] Boyce, J. (2020) *Imperial Mud: The fight for the Fens*, London: Icon Books
[4] www.gov.uk/government/news/new-chair-to-lead-task-force-on-sustainable-farming-of-peatlands
[5] Baseline Report (2020) Future Fens Flood Risk Management (December), www.ada.org.uk/wp-content/uploads/2021/05/Future-Fens-Flood-Risk-Management-Baseline-Report-Final_web.pdf
[6] Boyce, J. (2020) *Imperial Mud: The fight for the Fens*, London: Icon Books
[7] www.gov.uk/government/news/new-chair-to-lead-task-force-on-sustainable-farming-of-peatlands
[8] Environment Agency (2020) *National Flood and Coastal Erosion Risk Management* (July), www.environment-agency.gov.uk; NFU (2019) *Food and Farming in the Fens*, www.nfuonline.com/
[9] Environment Agency (2020) *National Flood and Coastal Erosion Risk Management* (July), www.environment-agency.gov.uk; NFU (2019) *Food and Farming in the Fens*, www.nfuonline.com/
[10] www.gov.uk/government/news/new-chair-to-lead-task-force-on-sustainable-farming-of-peatlands

11 Defra/Government Statistical Service (2019) *The Future Farming and Environment Evidence Compendium*, https://defra.gov.uk

12 Climate Change Committee (2020) Holme Fen. The Sixth Carbon Budget: Agriculture and land use; land use change and forestry. Holme Fen, www.theccc.org.uk/wp-content/uploads/2020/12/Sector-summary-Agriculture-land-use-land-use-change-forestry.pdf

13 Environment Agency (2020) *National Flood and Coastal Erosion Risk Management* (July), www.environment-agency.gov.uk; NFU (2019) *Food and Farming in the Fens*, www.nfuonline.com/

14 Slow the Flow (2020) www.slowtheflow.netwww.nationaltrust.org.uk/hardcastle-crags/features/natural-flood-management

15 Slow the Flow (2020) www.slowtheflow.netwww.nationaltrust.org.uk/hardcastle-crags/features/natural-flood-management

16 Climate Change Committee (2020). Holme Fen. The Sixth Carbon Budget: Agriculture and land use; land use change and forestry. Holme Fen, www.theccc.org.uk/wp-content/uploads/2020/12/Sector-summary-Agriculture-land-use-land-use-change-forestry.pdf

17 Climate Change Committee (2020). Holme Fen. The Sixth Carbon Budget: Agriculture and land use; land use change and forestry. Holme Fen, www.theccc.org.uk/wp-content/uploads/2020/12/Sector-summary-Agriculture-land-use-land-use-change-forestry.pdf

18 Boyce, J. (2020) *Imperial Mud: The fight for the Fens*, London: Icon Books

19 Boyce, J. (2020) *Imperial Mud: The fight for the Fens*, London: Icon Books

20 Defra/Government Statistical Service (2019) *The Future Farming and Environment Evidence Compendium*, https://defra.gov.uk

21 Environment Agency (2020) *National Flood and Coastal Erosion Risk management* (July), www.environment-agency.gov.uk

22 Baseline Report (2020) Future Fens Flood Risk Management (December), www.ada.org.uk/wp-content/uploads/2021/05/Future-Fens-Flood-Risk-Management-Baseline-Report-Final_web.pdf

23 Denver Sluice (2021) The Ouse Washes, www.ousewashes.info/sluices/denver-sluice.htm

24 Pollard, D. (2020) Interview with author

25 Baseline Report (2020) Future Fens Flood Risk Management (December), www.ada.org.uk/wp-content/uploads/2021/05/Future-Fens-Flood-Risk-Management-Baseline-Report-Final_web.pdf

26 Hall, J. (2021) Interview with author

27 Environment Agency (2020) *National Flood and Coastal Erosion Risk Management* (July), www.environment-agency.gov.uk

28 www.nationaltrust.org.uk/wicken-fen-nature-reserve

29 Haugh, R. (2014) *East Coast Storm Surge: what happened next?* www://bbc.co.uk/news/uk-england-norfolk-30183045 (December)

30 Environment Agency (2020) *National Flood and Coastal Erosion Risk Management* (July), www.environment-agency.gov.uk; NFU (2019) *Food and Farming in the Fens*, www.nfuonline.com/

31 Wall, T. (2019) This is a wake up call: the villagers who could be Britain's first climate refugees, 18 May, www.theguardian.com/environment

32 Hall, J. (2021) Interview with author

33 Town and Country Planning Association, www.tcpa.org.uk

34 Hetherington, P. (2015) *Whose Land is Our Land: the use and abuse of Britain's forgotten acres*, Bristol: Policy Press; Foresight (2010) *Land Use Futures: Making the most of our land in the 21st century*, Government Office for Science

35 NAO (2014) Strategic flood risk management, 5 November, www.nao.org.uk/report/strategic-flood-risk-management-2

36   Hetherington, P. (2015) *Whose Land is Our Land: the use and abuse of Britain's forgotten acres*, Bristol: Policy Press; Foresight (2010) *Land Use Futures: Making the most of our land in the 21st century*, Government Office for Science

37   Town and Country Planning Association, www.tcpa.org.uk

38   Bevan, J. (2019) www.gov.uk/government/speeches/escaping-the-jaws-of-death-ensuring-enough-water-in-2050 (March)

39   Environment Agency (2020) *National Flood and Coastal Erosion Risk Management* (July), www.environment-agency.gov.uk; NFU (2019) *Food and Farming in the Fens*, www.nfuonline.com/

40   Patient, S. (2021) Interview with author

41   Slow the Flow (2020) www.slowtheflow.netwww.nationaltrust.org.uk/hardcastle-crags/features/natural-flood-management

42   Monbiot, G. (2015) www.theguardian.com/commentisfree/2015/dec29/deluge-farmers-flood-grouse-moor-drain-land, 29 December

43   Carrington, D. (2019) Convert half of UK farmland to nature, urges top scientist www.theguardian.com/environment/2019/dec31/convert-farmland-to-nature-climate-crisis, 31 December

44   Climate Change Committee (2020) *Agriculture and Land Use, Land Use Change and Forestry*. Sixth Carbon Budget

45   Defra/Government Statistical Service(2019) *The Future Farming and Environment Evidence Compendium*, www.defra.gov.uk

Chapter 7

1   Socrates

2   Fferm Ifan (2020) https://fb.watch/3J2zsyrVGG/, 2 November

3   Office for National Statistics (2019) UK natural capital: peatlands, 22 July, https://www.ons.gov.uk/economy/environmentalaccounts/bulletins/uknaturalcapitalforpeatlands/naturalcapitalaccounts

4   Office for National Statistics (2019) UK natural capital: peatlands, 22 July, https://www.ons.gov.uk/economy/environmentalaccounts/bulletins/uknaturalcapitalforpeatlands/naturalcapitalaccounts

5   Office for National Statistics (2019) UK natural capital: peatlands, 22 July, https://www.ons.gov.uk/economy/environmentalaccounts/bulletins/uknaturalcapitalforpeatlands/naturalcapitalaccounts

6   https://cairngormsconnect.org.uk

7   Pearsall, W.H. (1950) *Mountains and Moorlands*, Collins New Naturalist Library

8   www.nts.org.uk/visit/places/mar-lodge-estate

9   Scottish Land Commission, Annual Business Plan 2020–21, www.landcommission.gov.scot

10   www.rewildingbritain.org.uk/explore-rewilding/what-is-rewilding/defining-rewilding

11   Jones, P. and Comfort, D. (2020) *Rewilding Ventures in the UK*, Town and Country Planning Association, https://tcpa.org.uk

12   Wentworth, J. and Alison, J. (2016) *Rewilding and Ecosystem Services*, Parliamentary Office of Science and Technology

13   Tree, I. (2018) *Wilding: The return of nature to a British farm*, London: Picador

14   Tree, I. (2018) *Wilding: The return of nature to a British farm*, London: Picador

15   Rebanks, J. (2020) *English Pastoral: An inheritance*, London: Allen Lane

16   Juniper, T. and Driver, A. (2021) Interviews with author (March, January)

17   Juniper, T. and Driver, A. (2021) Interviews with author (March, January)

18   https://wildland.scot/conservation

19 https://communitylandscotland.org.uk

20 Jones, P. and Comfort, D. (2020) *Rewilding Ventures in the UK*, Town and Country Planning Association, https://tcpa.org.uk

21 Pearsall, W.H. (1950) *Mountains and Moorlands*, Collins New Naturalist Library

22 https://businesswales.gov.wales/walesruralnetwork/local-action-groups-and-projects/projects/fferm-ifan-ecosystem-service-improvement-project

23 Fferm Ifan (2020) https://fb.watch/3J2zsyrVGG/, 2 November

24 Baird, A. (2021) Interview with author (January)

25 https://whoownsengland.org/2018/08/12revealed-the-aristocreats-and-the-city-bankers-who-own-englands-grouse-moors/

26 www.theguardian.com/environment/2018/aug/12/michael-gove-accused-of-letting-wealthy-grouse-moor-owners-off-the-hook

27 RSPB (2020) AGM position statement on grouse shooting, www.rspb.org.uk

28 Holden, P. and Tree, I. (2020() Podcast episode 11, www.sustainablefoodtrust.org.uk

29 Fferm Ifan (2020) https://fb.watch/3J2zsyrVGG/, 2 November

30 https://ons.gov.uk/economy/environmentalaccounts/bulletins/UKnaturalcapitalforpeatlands/naturalcapitalaccounts#main-points

31 https://cairngormsconnect.org.uk

32 Hunter, J., Peacock, P., Wightman, A. and Foley, M. (2013) *Towards a Comprehensive Land Reform Agenda for Scotland*, Scottish Affairs Select Committee, www.parliament.uk; www.gov.scot/publications/community-ownership-scotland-2018/

33 https://ons.gov.uk/economy/environmentalaccounts/bulletins/UKnaturalcapitalforpeatlands/naturalcapitalaccounts#main-points

34 Hunter, J., Peacock, P., Wightman, A. and Foley, M. (2013) *Towards a Comprehensive Land Reform Agenda for Scotland*, Scottish Affairs Select Committee, www.parliament.uk; www.gov.scot/publications/community-ownership-scotland-2018/

35 Natural Resources Wales (2020) https://naturalresources.wales/about-us/news-and-events/news/nrw-led-tackforce-set-to-accelerate-a-green-recovery-in-wales??land=en

36 Dickie, M. (2021) Duke's land sale provides a pathfinder for reforms, www.ft.com, 2 January

37 Hetherington, P. (2015) *Whose Land is Our Land: the use and abuse of Britain's forgotten acres*, Bristol: Policy Press

38 Dickie, M. (2021) Duke's land sale provides a pathfinder for reforms, www.ft.com, 2 January

39 https://bbc.co.uk/news/uk-scotland-47803110

40 Cunningham, R. (2020) Interview with author

41 www.nts.org.uk/visit/places/mar-lodge-estate

42 Povlsen, A.H. (2021) www.bbc.co.uk/sounds/play/m000rt8s

43 www.pressandjournal.co.uk/fp/news/highlands/2022736/sutherland-couples-disappointment-as-billionaire-snaps-up-their-dream-house/

44 Wilson, B. (2021) www.heraldscotland.com/news/19005143.brian-wilson-want-revitalise-rural-scotland/?=erec

45 Hunter, J. (2019) *Repeopling Emptied Places: Centenary reflections on the significance and the enduring legacy of the Land Settlement (Scotland) Act 1919*, www.landcommission.gov.scot

46 Wilson, B. (2021) www.heraldscotland.com/news/19005143.brian-wilson-want-revitalise-rural-scotland/?=erec

47 Brooks, L. (2016) A new dawn for land reform in Scotland? www.theguardian.com, 17 March

48 Cunningham, R. (2020) Interview with author

49 https://cairngormsconnect.org.uk

50 Hetherington, P. (2015) *Whose Land is Our Land: The use and abuse of Britain's forgotten acres*, Bristol: Policy Press, pp 14–16

51 Hetherington, P. (2015) *Whose Land is Our Land: The use and abuse of Britain's forgotten acres*, Bristol: Policy Press, pp 14–16

52 Hetherington, P. (2015) *Whose Land is Our Land: the use and abuse of Britain's forgotten acres*, Bristol: Policy Press

53 Young, B. (2021) What Whitehall might learn from Scotland and Wales, *Journal of the Town and Country Planning Association* (April) www.tcpa.org.uk

Chapter 8

1 Isiah 32:18

2 Gibson, R. (2020) *Reclaiming Our Land*, Highland Heritage, Rob Gibson in association with Highland Heritage Educational Trust

3 Burn-Murdoch, J. (2017) Small towns left behind as exodus of youth to cities accelerates, www.ft.com/content/2312924c-ce02-11e7-b781-794ce08b24dc

4 Pragmatix Advisory (2020) *Rural Recovery and Revitalisation*, Report for CPRE, English Rural, and Rural Services Network

5 Pragmatix Advisory (2020) *Rural Recovery and Revitalisation*, Report for CPRE, English Rural, and Rural Services Network

6 Defra (2020) *Rural Housing; additions to affordable housing stock*, www.gov.uk

7 Defra (2020) *Rural Housing; additions to affordable housing stock*, www.gov.uk

8 Sky News (2018) Rural housing in England, New Lines, www.news.sky.com

9 Hetherington, P. (2015) *Whose Land is Our Land: The use and abuse of Britain's forgotten acres*, Bristol: Policy Press

10 CRHA and Community Land Trusts in Cornwall, crha.org.uk, St Minver Community Land Trust

11 Halifax (2021) Halifax Seaside Town Review, www.lloydsbankinggroup.com, 27 May

12 Hetherington, P. (2015) *Whose Land is Our Land: The use and abuse of Britain's forgotten acres*, Bristol: Policy Press

13 Hetherington, P. (2015) *Whose Land is Our Land: The use and abuse of Britain's forgotten acres*, Bristol: Policy Press

14 Sky News (2018) Rural housing in England, New Lines, www.news.sky.com

15 Shucksmith, M. (2018) *Rural Policy after Brexit*, Academy of Social Sciences, Routledge, https://doi.org/10.1080/21582041.2018.1558279

16 Newby, H. (1980) *Green and Pleasant Land: Social change in rural England*, Harmondsworth: Penguin

17 www.westharristrust.org

18 www.armadalefarm.co.uk

19 Cunningham, R. (2020) Interview with author

Chapter 9

1 Ingersoll, R. (1833–99) Courtesy of the late Professor Philip Lowe

2 Fferm Ifan, www.nationaltrust.org.uk/features/fferm-ifan

3 Landlines (2017) *Why We Need a Strategic Approach to Land*. CPRE, the countryside charity (March), www.cpre.org.uk; www.gov.uk/government/news/farmers-invited-to-take-first-step-towards-greener-future; www.ft.com/content/e33d5f12-f62-11e7--8715-e94187b3017e

4   Booth, R. (2021) *Johnson's planning laws an 'utter disaster' say countryside campaigners*, 11 May, www.theguardian.com/politics/2021/may/11/johnsons-planning-laws-an-utter-disaster-say-countryside-campaigners

5   Eustice, G. (2021) *Sustainable Farming Incentive: Defra's plans for piloting and launching the scheme* (March), www.defra.gov.uk; Boyd, I. (2019) https://theconversation.com/climate-crisis-the-countryside-could-be-our-greatest-ally-if-we-can-reform-farming-126304

6   Landlines (2017) *Why We Need a Strategic Approach to Land*, CPRE, the countryside charity (March), www.cpre.org.uk; www.gov.uk/government/news/farmers-invited-to-take-first-step-towards-greener-future; www.ft.com/content/e33d5f12-f62-11e7--8715-e94187b3017e

7   Landlines (2017) *Why We Need a Strategic Approach to Land*, CPRE, the countryside charity (March), www.cpre.org.uk; www.gov.uk/government/news/farmers-invited-to-take-first-step-towards-greener-future; www.ft.com/content/e33d5f12-f62-11e7--8715-e94187b3017e

8   Booth, R. (2021) *Johnson's planning laws an 'utter disaster' say countryside campaigners*, 11 May, www.theguardian.com/politics/2021/may/11/johnsons-planning-laws-an-utter-disaster-say-countryside-campaigners

9   Landlines (2017) *Why We Need a Strategic Approach to Land*, CPRE, the countryside charity (March), www.cpre.org.uk; www.gov.uk/government/news/farmers-invited-to-take-first-step-towards-greener-future; www.ft.com/content/e33d5f12-f62-11e7--8715-e94187b3017e

10 Lang, T. (2020) *Feeding Britain: Our food problems and how to fix them*, London: Pelican

11 Defra/Government Statistical Service (2019) *The Future Farming and Environment Evidence Compendium*, May, www.defra.gov.uk

12 Tett, G. (2020) Why we need to put a number on our natural resources, 6 September, www.ft.com/content/e82d6703-27d1-4bf0-972e-16fa726811d0

13 Eustice, G. (2021) *Sustainable Farming Incentive: Defra's plans for piloting and launching the scheme* (March), www.defra.gov.uk; Boyd, I. (2019) https://theconversation.com/climate-crisis-the-countryside-could-be-our-greatest-ally-if-we-can-reform-farming-126304

14 Stevenson, D. (2021) Forestry investors see the wood for the trees, 16 August, www.ft.com/content/2a4910ae-d8d0-426e-a1be-06d725c4d6fa

15 www.nffn.org.uk

16 www.agricology.co.uk/field/farmer-profiles/martin-lines

17 Lang, T. (2020) *Feeding Britain: Our food problems and how to fix them*, London: Pelican

18 Begg, P. (2020) Interview with author

19 Prentis, V. (2021) Interview on *Farming Today*, BBC Radio 4, 13 March; www.bbcnews.com

20 Prentis, V. (2021) Interview on *Farming Today*, BBC Radio 4, 13 March; www.bbcnews.com

21 Carrington, D. (2020) Future looks bleak for the soil that underpin life on Earth, warns UN, *The Guardian*, www.theguardian.com

22 Booth, R. (2021) *Johnson's planning laws an 'utter disaster' say countryside campaigners*, 11 May, www.theguardian.com/politics/2021/may/11/johnsons-planning-laws-an-utter-disaster-say-countryside-campaigners

23 Fferm Ifan. www.nationaltrust.org.uk/features/fferm-ifan

24 Tenant Farmers' Association (2021) Strong interest in BPS lump sum exit scheme, www.tfa.org.uk, February

25 Defra/Government Statistical Service (2019) *The Future Farming and Environment Evidence Compendium*, May, www.defra.gov.uk

26 Defra/Government Statistical Service (2019) *The Future Farming and Environment Evidence Compendium*, May, www.defra.gov.uk

27 Defra/Government Statistical Service (2019) *The Future Farming and Environment Evidence Compendium*, May, www.defra.gov.uk

28 NAO (2019) Report: Early review of the new farming programme (Defra) HC 2221, Session 21, 2017–19, June, www.nao.org.uk

29 Clark, C., Scanlon, B. and Hart, K. (2019) *Less Is More*, RSPB, Wildlife Trusts, National Trust, www.nationaltrust.org.uk

30 Kynaston, D. (2007) *Austerity Britain 1945–51*, London: Bloomsbury

31 Agyemang, E. (2021) The future of UK inheritance tax – lessons from other countries, https://www.ft.com/content/c52faece-2e07-4ae3-912d-83cae264093e, 28 May

Index

References to figures and photographs appear in *italic* type.

supermarkets
    farm subsidy beneficiaries 18, 19, 42
    food pricing control 48
    LSA crops contract 34
    supply chain fragility 30, 46, 49
Sustainable Farming Incentive (SFI) 135, 141
Sustainable Food Trust (SFT) 107

T

Tax Justice Network 57
Taylforth, Eric 72–3
Taylforth, Sue 72
Tebay Services, diversification venture 67–9
Tenant Farmers' Association (TFA) 55, 123
tenant farming
    climate change issues 55
    corporate interest 56–7, 58
    landlords' financial privileges 56–8
    self-redundancy issues 10, 123, 143
    subsidy changes, viability concerns 10, 20, 24, 55, 143
Thatcher, Margaret 21, 125
Tolhurst, Ian 39, 40, 41, 48, 49, 133, 138
Town and Country Planning Act 1947 7, 36, 113, 135
Town and Country Planning Association (TCPA) 92–3, 94
trade deals, agricultural 10, 22, 36
Tree, Isabella 101, 107
Trevelyan, Charles, Sir 32
Tudge, Colin 49
Tudge, Ruth 49

U

Uig, Outer Hebrides 109
upland farming
    biodiversity debate 20, 62, 65, 71–3, 120
    de-stocking, ELMS led agenda 72–3, 75
    diversification, Tebay Services 67–9
    forestation and carbon trading 69–70

North Pennies 75
sustainable diversification 70–1, 79–80, 96
viability, potential and concerns 8, 10–11, 17, 68, 79, 143–4
wool production's decline 73–4, 76–8

V

Vermuyden, Cornelius 85
vertical farming 19, 52–3, 82, 108

W

Walker, Peter 34
Wear, Andy 76
West Harris 123
Whitehall Farm, Peterborough 27, 28, 32, 62
Wicken Fen 88
Wild Ennerdale, Lake District 103
Wildland 114, 115–16, 132
Wildlife Trust 104–5
Williams, Raymond 12
Williams, Tom 136
Williamson, Richard 61
Wilson, Brian 13, 117
Wilson, Ross 8, 9, 10–11, 14, 17, 24
Wise, M.J. 34
wool industry
    decline factors 76–8
    farmer complaints and critics 73, 74, 76, 77
    Harris Tweed Act 1993 78
    history 74
    revival advocates 76–7, 78–9
Wooler, Northumberland 126, 130

Y

Yorkshire Water 93
Young, Barbara, Baroness 121, 136

Z

Zephaniah, Benjamin 11